Flowering Trees

Horticulture Science Series - 12

NIPA® GENX ELECTRONIC RESOURCES & SOLUTIONS P. LTD.
New Delhi-110 034

Other Volumes of the Series

HORTICULTURE SCIENCE SERIES

Series Editor : Prof. K.V. Peter

Former Vice Chancellor and Presently Professor of Horticulture
Kerala Agricultural University, Thrissur - 680 656, Kerala

1. Aromatic Plants
   Baby P. Skaria, P.P. Joy, Samuel Mathew,
   Ancy Joseph & Ragina Joseph
2. Medicinal Plants
   Kurian Alice & M. Asha Shankar
3. Fruit Crops
   T. Radha & L. Mathew
4. Vegetable Crops
   Gopalakrishnan, T.R.
5. Spices
   E.V. Nybe and N. Mini Raj & K.V. Peter
6. Propagation of Horticultural Crops
   Rajan S. & B.L. Markose
7. Post Harvest Technology of Horticultural Crops
   Sudheer K.P. & V. Indira
8. Commercial Crops Technology
   Kurian Alice & K.V. Peter
9. Tuber and Root Crops
   Palaniswami P. & K.V. Peter
10. Flowers for Trade
    V.L. Sheela
11. Management of Horticultural Crops: (*in 2 parts*)
    Kumar Pradeep, T. Suma, B. Jyothibhaskar,
    Satheesan K.N. & K.V. Peter

Horticulture Science Series Vol. 12

Series Editor:
Prof. K.V. Peter, Former VC and Presently Professor of Horticulture Kerala Agricultural University, Thrissur - 680 656, Kerala

FLOWERING TREES

Authors

P.K. VALSALAKUMARI, Professor

P.K. RAJEEVAN, Professor & Head

P.K. SUDHADEVI, Professor

C.K. GEETHA, Professor

Department of Pomology and Floriculture, College of Horticulture
Kerala Agricultural University, Thrissur - 680 656 (Kerala)

NIPA® GENX ELECTRONIC RESOURCES & SOLUTIONS P. LTD.
New Delhi-110 034

NIPA® GENX ELECTRONIC
RESOURCES & SOLUTIONS P. LTD.

101,103, Vikas Surya Plaza, CU Block
L.S.C. Market, Pitam Pura, New Delhi-110 034
Ph : +91 11 27341616, 27341717, 27341718
E-mail: newindiapublishingagency@gmail.com
Web: www.nipabooks.com

For customer assistance, please contact
Phone: + 91-11-27 34 17 17
Fax: + 91-11- 27 34 16 16
E-Mail: feedbacks@nipabooks.com

ISBN: 978-81-19103-74-4

Composed and Designed by NIPA®.

Dedicated to

All those who love
The great art and science of Horticulture

National Academy of Agricultural Sciences

Prof. M.S. Swaminathan
President

NASC Complex
DPS Marg, P.O. Pusa
New Delhi - 110 012

Tel. 91-11-25846055
91-11-25846051
91-11-25846052
91-11-25846053
Fax 91-11-25846054
E-mail naas@vsnl.com
Website
//education.vsnl.com/naas
P.O. Box No. 11325

Foreword to the Series

A second green revolution aimed to meet the food requirements of about 1.8 billion people in India by 2010 is on the anvil. It is, however, imperative that the attainment of the food and nutritional security of this vast population should be combined with the preservation of clean environment. Horticulture encompassing fruits, vegetables, ornamentals, plantation crops, spices, tuber crops, medicinal and aromatic plants, besides harvest and post harvest technologies has a special role in providing food and nutritional security, rural poverty alleviation, employment generation as well as improving the quality of life in our country sides.

India, incidentally, is the world's leading producer of fruits and is followed by Brazil; in vegetables it occupies the second place after China. India is also the land of spices producing an array of such crops. Indeed, the varied eco-climatic conditions prevailing in different parts of this country make it possible to grow a disparate range of fruits, vegetables and ornamental plants at any time of the year.

With the opening up of the world markets under the WTO regime, however, international competition for Indian horticultural products has reached unprecedented levels, necessitating strategies for quality control, branding and product development. In particular, production of organic vegetables, fruits, cut flowers, pepper, ginger, tea, coffee, rubber and cardamom has of late become demand-driven.

Among the inputs for higher productivity in horticultural crops, seeds and quality planting materials account for 20 to 30% of the potential productivity. Transgenic crops are a reality now; yet the technology for production of transgenics is a far cry in many situations. Furthermore, water is a limiting factor in several environments. Hence, water saving crops and those with high water use efficiency should be preferred over other crops and management practices. Likewise, the use of biodegradable plastics, biological control of pests and diseases, and new product development are receiving increasing attention, especially in the arena of horticultural crop management. Marketing, village knowledge centres, packaging industry, zero-energy cool chambers and a variety of economically viable technologies, however, remain to be streamlined.

The present series on **Horticulture Sciences,** edited by Professor K.V. Peter, Vice-Chancellor, Kerala Agricultural University, is a good compilation of the scientific knowledge and experience gained over the past 30 years in this important sector of Indian economy. Each volume in the series is authored by accomplished researchers such as Dr. T. Radha and Dr Lila Mathew (*Fruit Crops*), Dr. T.R. Gopalakrishnan (*Vegetables*), Dr. Alice Kurian and Dr. K.V. Peter (*Commercial Crops Technology*), Dr. Alice Kurian and Dr. M. Asha Sankar (*Medicinal Plants*), Dr. Baby P. Skaria (*Aromatic Plants*), Dr. P. Palaniswami (*Tuber and Root Crops*), Dr. E.V. Nybe and Dr N. Mini Raj (*Spices*), Dr. S. Rajan and Dr Baby Lizy Markose (*Propagation of Horticultural Crops*), Dr. K.P. Sudheer and Dr. V. Indira (*Post Harvest Technology of Horticultural Crops*), Dr. V.L. Sheela (*Flowers for Trade*), Dr. K. Pradheep (*Management of Horticultural Crops*), Dr. V. Kanthaswamy (*Coastal Horticulture*), Dr. Sabina George (*Ornamentals in Garden*) and Professor K.V. Peter (*Basics of Horticulture*). The New India Publishing Agency, New Delhi deserves special credit for publishing this useful series, especially in view of the fact that the Government of India is currently implementing the National Horticultural Mission.

Prof. M.S. Swaminathan, FNA, FRS
President, National Academy of Agricultural Sciences
and Chairman, National Commission on Farmers
New Delhi - 110 012

Indo-American Hybrid Seeds (India) Pvt. ltd.

Speciality
Hybrid Flower & Vegetable Seeds

SYMBOL OF QUALITY

Registered Office:
Post Box No. 7099
17th Cross, II "A" Main
K.R. Road, Banashankari II Stage
Bangalore-560 070. India
Phone: 26760111 Fax: 91-80-26761479
Website : www.indamseeds.com

FOREWORD

Nature has gifted man with trees for his sustainable livelihood. Trees provide him with food, shelter, fuel, drugs, materials like paper for industry, protect from environmental hazards like pollution and soil erosion, provide a comfortable microclimate apart from beautifying the environment. Sri. Aurobindo described trees as beautiful symbol of nature's aspiration towards light.

Growing and worshipping trees were associated with Indian culture from ancient period. Place of trees in environmental maintenance and food chain was well understood in those days. Saints and ancient philosophers encouraged people to 'grow more trees' which we have rediscovered and trying to practice now. Fortunately, India has rich wealth of ornamental trees, both indigenous and introduced from different parts of world by kings and rulers of respective periods.

Trees are integral part of any landscape. Apart from beautifying surroundings trees are used functionally to improve the quality of environment particularly in urban areas, where the environment is degrading at a faster rate. Trees are used widely in monitoring and controlling environmental pollution. The towns and villages are healthier, more beautiful and more efficient by well-planned streets with avenue planting.

Trees can play an important role in modern landscaping by reducing glare, screening, framing views, as physical barriers, to control solar radiation, temperature, wind and humidity. Their ecological value in noise, dust and erosion control is well recognized.

Studies in environmental psychology proved beneficial effects of trees in human health and well being. Trees are recommended for planting in industrial areas, hospitals, schools and colleges, offices, star hotels, religious places, home gardens and in cities and villages wherever space is available.

Trees are selected for specific uses based on the purpose of growing, tree architecture, colour of foliage or flower, availability of space, soil and climatic factors. Native trees and leguminous trees, which enrich soil with nitrogen, are gaining importance in modern landscaping and bioaesthetic planning of cities and villages.

Selection of planting material, suitable site, proper installation and tree care with special emphasis on pruning are the major factors which enable trees to fulfil their functional and aesthetic attributes in the landscape for benefit of mankind.

All the above aspects about flowering trees are described in this book. It also contains relevant information from '*Vrikshayurveda*', the ancient 'Science of plant life' written by Surapala. Descriptions of indigenous and other ornamental trees with reference to their landscaping uses are given in this book.

A flowering tree is a joy forever. As rightly suggested in *Vrikshayurveda* trees should be carefully nurtured as through their shade, flowers and fruits they help immensely in the pursuit of *dharma*, *artha* and *kama*. I congratulate Dr P K Valsalakumari Professor, Kerala Agricultural University, Thrissur for authoring the book. Dr Valsalakumari is a renowned teacher and a researcher on ornamental trees. She has handled the topic in a lucid style which will be appreciated by one and all. Prof. K V Peter, Former Vice-Chancellor, Kerala Agricultural University has edited the Flowering Trees. I also compliment New India Publishing Agency, New Delhi for selecting the best authors available.

Manmohan Attawar
Chairman
Indo-American Hybrid Seeds
Bangalore

PREFACE

Landscaping is the art and science of achieving a total visual effect in an area to give it a sense of wholeness in the background of topography, buildings and the surrounding environment. The objective of modern landscaping is not merely to beautify the environment. It emphasizes the functional use of plants so as to modify the microclimate of men's habitat and the area surrounding it and ultimately to produce a landscape that would be self sustaining following the principle of 'energy saved is energy gained'. The term 'Environscaping' is of much relevance in the present context.

Trees form the back bone of any landscape. Apart from the aesthetic value provided by their flowers and foliage, trees are looked upon for their functional values as well. Climate control is the most important function of trees in the landscape which they provide naturally. Trees reduce temperature around the area inhabited by them by absorbing solar radiation. Trees help to moderate the *heat island*, a condition formed due to significant increase in temperature.

In industrial areas trees serve as effective tools for pollution control. For every ton of new wood that grows in a tree, 1.8 tons of carbon dioxide is removed from the air and 1.3 tons of oxygen is produced. Air pollution control is the most important way that trees improve the environment. On an average, trees remove roughly one-half the total weight of air pollutants.

Trees also reduce noise pollution from all sources. Trees and their associated ground cover near water streams, rivers and lakes serve as a sponge, reducing the flow of pollutants such as agricultural chemicals before they reach the fresh water. Trees can cut energy costs by reducing the temperature in and around buildings and reduce air-conditioning costs. In this way, trees indirectly reduce carbon dioxide emissions. Trees and other vegetation protect the soil from erosion. Tree canopy reduces the force of the falling rain, thus preventing the direct impact of rain drops on soil.

Soil and water quality are of prime importance for biodiversity development, which is greatly favoured by trees. Leaves and other litter that accumulate under trees create an environment conducive and ideal for microorganisms, earthworms and other organisms that help maintain soil porosity and also enrich the soil fertility. Trees also improve the habitat development and facilitates sheltering place for epiphytes like orchids, lichens, fungi, algae, insects, birds, reptiles and other organisms.

The most recent use of trees is as energy crops. Many tree species are identified as potential sources of biomass for the production of alternate energy. It offers tremendous environmental advantages compared with coal, petroleum and natural gas. It is renewable and as it grows biomass removes carbon dioxide from the air, thus helping to slow down the build up of 'greenhouse gases' that threatens to warm the globe and alter earth's climate for ever.

Worship of the trees is an ancient Indian practice. The ancient Aryans worshipped nature. Trees were always revered and several rituals were connected with them. Trees were given a place next to God in Hindu Mythology. Majority of the temples in India have huge trees or sacred groves in the vicinity. Most these temple trees are noted for their medicinal properties and sacred groves are a well known source of Ayurvedic medicines. Recently scientists at the M.S. Swaminathan Foundation for Research found that trees in mangroves could cure several diseases.

Many trees were considered sacred. The banyan tree, peepal tree, ashoka tree, bhang tree, neem tree, red sandalwood tree, mango tree, coconut tree and bamboo were considered as sacred and legendary trees of India. The Banyan tree (*Ficus benghalensis*) is the National tree of our country. Its branches root themselves like new trees over a large area. The roots then give rise to more trunks and branches. Because of these characteristics and its longevity, the tree is considered immortal and is an integral part of the myths and legends of India. The city of Vadodara in western India is named after banyan tree (*Vadavriksha*).

The tree, with its rising branches to the sky symbolizes nature's aspiration for light, as described by Sri Aurobindo. It gives shade even to a person who cuts it down. It gives its sweet delicious fruits to the person who harms it. When the tree stops yielding the trunk is made into furniture which again is useful to man. Our great nuclear scientists Dr. Homi J. Bhabha and Dr. A.P.J. Abdul Kalam were known as 'Friends of trees'. By protecting trees and plants we are protecting and preserving nature. We should always remember that *we have not inherited this earth, but have borrowed it from the generations to follow us.*

CONTENTS

1

India's Heritage of Ornamental Trees

Trees rise towards the sky, beautiful symbol of Nature's aspiration for light. Sri Aurobindo writes "When the sun sets and all become silent, sit down under a tree for a moment and put yourself into communication with Nature: you will feel rising from the earth, from below roots of the trees and mounting upwards and coursing through their fibres upto the highest outstretching branches, the aspiration of an intense love and longing, - a longing for something that brings light and gives happiness. There is a yearning so pure and intense that if you can feel the movement in the trees, your own being too will go up in an ardent prayer for the peace and light and love that are unmanifested here".

Plants have more of it in their physical being than in men. Their whole life is a worship of light. Light is of course the material symbol of the Divine, and the sun represents, under material conditions, the Supreme Consciousness. The plants have felt it quite distinctly in their own simple and blind way. Their aspiration is intense, if you know how to become aware of it. The plants are very psychic, but they can express it only by silence and beauty.

In ancient Indian culture, love of plants and flowers was considered as a valuable help for finding and uniting with the psychic. May be because of this reason, India has a rich heritage in ornamental gardening and especially in growing trees. Trees and gardens have always been associated with Indian culture, from the ancient vedic times. We get beautiful descriptions of plants and gardens from the Vedas, ancient Hindu epics, Sanskrit literature, sculptures and paintings.

It is proven beyond doubt that the Science of plant life was developed in India even prior to the 6th century. *Vrikshayurveda* written by Surapala is an ancient Sanskrit text on the science of plant life. The place of tree in environmental maintenance and food chain was well understood in those days. The saints and philosophers of the time extolled the people to 'grow more trees'- a slogan that we seem to have rediscovered during the past few decades.

Trees in Hindu epics and ancient Sanskrit literature

Peepal tree (*Ficus religiosa*) is mentioned in vedas. In Ramayana, we get a detailed description of 'Asokavana' in which Sita was held captive. Asoka tree (*Saraca indica*) was predominant in the garden. Several other trees like *Terminalia arjuna, Mesua ferrea, Ficus benghalensis, F. religiosa, Michelia champaca, Butea monosperma* and *Cassia fistula* are mentioned in the Ramayana. Almost all of them have been described in the Mahabharata. A description of the layout of gardens and parks and artificial lakes in the city of Indraprastha is given in the Sabha-Parva of the Mahabharata. Lord Krishna's 'Vrindavanam' was full of trees and other plants. The association of Lord Krishna with the kadamba tree (*Anthocephalus cadamba*) is well known.

In the ancient Sanskrit works of Panini (*Ashtadhyayi*), several beautiful trees are mentioned, such as *Ficus spp.* (*F. religiosa, F. benghalensis, F. infectora*), *Butea monosperma, Prosopis juliflora, Anthocephalus cadamba* and a few others. The poet Asvaghosa described the Nandanavana in which Siddharatha Gauthama saw flowering trees and lotuses. Lord Buddha was born under a peepal tree in a garden. The Bodhi tree under which the Buddha attained nirvana, is sacred to the Buddhists. During the Buddhist period, gardens were laid out around the monasteries and 'Stupas' and there were beautiful gardens in ancient Universities of Nalanda and Taxila. It is said that the famous Japanese style of gardening originated in India, was taken by the Buddist monks to China, and from there to Japan.

The planting of roadside avenue trees (*margeshu vriksha*) was an important contribution of Emperor Ashoka (233 B.C). Similarly, Shudraka (100 B.C) gave an account of gardens and flowers in 'Mrichhakatikam'. Kalidasa (about 75 B.C.) gave romantic descriptions of trees and other plants in his famous plays and poems. In his play 'Abhignana Shakuntalam', he provided poetic descriptions of plant associations. He described the madhavilata (*Hiptage madablota*) creeping on a mango tree is like a beautiful girl joining with a handsome boy on the occasion of Sakuntala joining with Dushyanta. He described several beautiful trees like ashoka (*Saraca indica*), kadamba (*Anthocephalus cadamba*), arjuna (*Terminalia arjuna*), vakula (*Mimusops elengi*), palasha (*Butea monosperma*), parijata (*Nyctanthes arbor-tristis*) and kavidara (*Bauhinia variegata*). Beautiful descriptions of trees and other woody plants were given in Kalidasa's very famous work 'Meghasandesa'.

The art of gardening was described by Sarangadhara (A.D. 1300) in his 'Upavana Vinoda', and 'Sarangadhara Paddhati', wherein mention is also made of some trees. Vatsyayana (A.D. 300-400) rendered interesting accounts of four kinds of gardens, namely, paramadodyana, udyan, vrikshavatika and nandanavana. Classical Sanskrit literatures, Upavana Vinoda and Sarangadhara Paddhati as mentioned above, as well as the flower and tree motifs delineated in old sculptures and the architecture of Mathura (Kanishka period A.D. 100-600) bear testimony to the importance of gardening, trees and flowers in Indian culture.

Important native ornamental flowering trees, many of which are mentioned in ancient literature are kachnar (*Bauhinia variegata*), amaltas (*Cassia fistula*), pink cassia (*Cassia nodosa*), dhak or flame of the forest (*Butea frondosa*), Indian coral tree (*Erythrina indica*), pride of India (*Lagerstroemia flos-reginae*), scarlet cordia (*Cordia sebestena*), yellow silk cotton (*Cochlospermum gossypium*), karanj (*Pongamia glabra*), tulip tree or bhendi (*Thespesia populnea*), barna (*Crataeva roxburghii*), chalta (*Dillenia indica*), ashoka, kadamba and rhododendrons.

Vrikshayurveda - The ancient science of plant life

Vrikshayurveda written by Surapala was translated by Smt. Nalini Sadhale and was published by the Asian Agri-History Foundation, Andhara Pradesh in 1994. As stated by Surapala, the book was written after studying the various *sastras* of the ancient sages and after being convinced about their authenticity.

Vrikshayurveda, which means "The Science of Plant Life", mainly deals with various species of trees and their healthy growth and productivity. The text mentions about 170 species of plants, including herbs, shrubs, and trees. There are 325 systematically arranged verses, beginning with a salutation of Lord Ganesha, followed by glorification of trees, and composition on tree-planting and production. Various chapters deal with the raising of orchards, agri-horticulture, and tree planting in homesteads. Special references are made on procuring, preserving and treatment of seeds and planting materials; preparation of pits for planting; selection of land (soil); method of irrigation and ways to locate ground water; nourishment and fertilizers; diseases of plants and plant protection methods. It also describes use of plant species as indicators of crop and animal production and description of sacred plants and biodiversity conservation.

It seems that the text was written to help conserve plant biodiversity of selected tree species. Surapala elaborated on the importance and glory of trees and took advantage of common peoples' believes, religious attitudes, superstitions and folk knowledge to persuade people to plant trees. He also elaborated on the benefits that a person was likely to derive by planting trees, both in this life and after life (in heaven) or after re-birth.

Vrikshayurveda in contemporary times

In today's world, global environment issues relating to sustainable development have emerged as topics of major concern. We are now facing the challenge of looking beyond the obvious issues of producing more food together with giving a fuller recognition to issues of biodiversity. *Vrikshayurveda* is full of references that suggest that raising tree is a means of attaining the four broad aims of life: *dharma, artha, kama,* and *moksha,* which broadly mean a holistic development of mankind and its welfare in all the spheres including inter-generational equity. These are key words for today's sustainable development of agriculture.

The place of horticulture in ecology both at the home garden scale and at the field scale was well understood. Some of Surapala's statements which reveal the importance of trees are given below.

- Ten wells are equal to one pond. Ten ponds are equal to one lake. Ten lakes are equal to one son. Ten sons are equal to one tree.
- He who plants even a single *asvattha, (Ficus religiosa)* wherever it may be, as per the prescribed mode, goes to the abode of Hari.
- He who plants a couple of banayan tree (*Ficus benghalensis*) as per the prescribed mode would go to the abode of Siva and many heavenly nymphs will attend upon him.
- After planting neem trees (*Azadirachta indica*) a person well-versed in *dharma* attains the abode of the Sun Indeed! He resides there for a long period.
- By planting four *plaksha* trees (*Ficus lacor*) a person doubtlessly obtains the fruit of Rajasuya sacrifice.
- He who has planted *maduka* (*Madhuka indica*) has propitiated Parvati, has become free from diseases, and has worshipped all the deities.
- One should plant seven *palasa* trees (*Butea monosperma*) or even one. One attains the abode of Brahma and enjoys the company of gods by doing so.
- By planting all kinds of other trees, useful for fruits and flowers, a person gets a reward of thousand cows adorned with jewels.

Trees in the Mughal period

Like many crop plants, several of the trees now grown in our country, were introduced from abroad. These exotic trees came from Europe, America, Africa, China, Japan and other countries. It is said that most of them were introduced during the Mughal and British periods. The

Mughals were descendants of Timur, who ruled the Persian and Turkic worlds around the fifteenth century. Babur, the ruling descendant of Timur occupied the throne of Ferghana, northeast of Samarkand. After the loss of Ferghana and then Samarkand, Babur took Kabul in 1504. Well known as a poet and musician, Babur was also a talented designer who built numerous gardens, inspired by his love for the lush and beautiful cities of Samarkand and Herat.

Babur's last and most important conquest was Hindustan, and one of his first acts, when he arrived on the subcontinent to stay in 1526, was to construct gardens. He declared to make "that charmless and disorderly Hind (India)" more like home. Eventually the image of the garden became all pervasive in the Mughal world.

With the renaissance of gardening in India by the Mughal emperors beginning with Babur, many plant species were brought by them from Persia and Central Asia. Many of these are mentioned in autobiographies and other books written during those days. Besides, in the Mughal painting also we find illustrations of many trees and flowers. These were also used to illustrate the borders of those painting. In the book *Bagh-e-wafa*, the emperor Babur presented a description of gardening in India. The Mughal emperor introduced several types of plants, many of which were planted in Kashmir where the climate was more suitable for growth of such plants than that of the plains. The species brought by them included the famous chinar tree (*Platanum orientalis*), which is the most beautiful tree in Kashmir even today.

Mixed avenues of cypresses and flowering trees like *Bauhinia variegata*, orange and lemon, were planted. Oranges and pomegranates were used to border the square plots. For the Mughals, every plant had some symbolic background and the selection was done accordingly. For example, oranges, pomegranate, almonds, plums, and white *kachnars* are the symbols of youth and life, while cypresses symbolize death and senility.

Trees in the British period

During the British period many species of trees were introduced mainly by Englishmen and the Portuguese. These were mostly brought in by missionaries and priests, civil servants and individual amateur gardeners. One of the important missionaries who introduced a number of exotic plants was Dr. Firminger, an Englishman, who wrote a book on gardening, giving descriptions of various species of trees and other plants in the year 1863. The book entitled "Firminger's Manual of Gardening in India" is an authoritative reference book on ornamental flowering plants even today. The rich natural flora of India was documented during the British period. Several of our native flowers, particularly the attractive flora of the Himalayas including many alpine species, have been introduced to other countries.

Several botanical gardens were established in different agro climatic parts of the country to study the performance of the exotic plants. Any plant from anywhere in the world can find a place in the Indian subcontinent owing to an obvious advantage geographically. Though, very many plants of a good deal of importance have their original home on this land, due to a high degree of endemism their protection becomes even more important lest they be lost forever. Botanic Gardens in different phytogeograpical regions become a must for ensuring the preservation of flora, creating a sizeable natural resource and rearing endangered flora. Apart from their scientific and educational value, they provide an atmosphere of tradition and tranquility.

2

Spiritual and Religious Significance of Trees

The cultural history of India is associated with religious believes and any emphasis on arguments dealing with plant-man-God relationship appeals to people even today. This attitude prevails throughout India and many places of worship have continued to preserve *in situ* biodiversity of specific species of plants.

Religious importance of trees

Several verses of *Vrikshayurveda* elucidate the relationship between certain species of plants and the religious life of people. Specific plant species were considered sacred and especially associated with specific gods and goddesses like *Aegle marmelos* and *Ficus benghalensis* with Siva, *Azadirachta indica* with the Sun, *Mangifera indica* with Garuda, *Butea monosperma* with Brahma, *Ficus glomerata* with the Moon, *Madhuka indica* with Parvati and other deities, and *Ficus religiosa* with Vishnu (Hari).

Auspicious and inauspicious trees

Surapala provided a list of auspicious and inauspicious trees for planting and their location near a house, e.g., *Ficus benghalensis* is auspicious when planted in the east but not in the west; *Ficus glomerata* in south, not north; *Ficus religiosa* in the west, not east and *Ficus lacor* in the north, not south side of a house. He also mentioned that laying of a garden to the south, southwest, or northeast of a house leads to quarrels, distress, and adversity. Therefore, he recommended that gardens should be laid out on the western, northern, and eastern side of a house. He mentioned that shadows of trees should not fall on the house.

Surapala also listed several plant species considered inauspicious. The planting or natural occurrence of such plants leads to unhappiness (*Ricinus communis, Bauhinia variegata, Cordia wallichii, Terminalia arjuna, and Pongamia pinnata*), danger from enemies (thorny plants, in general), loss of wealth (*Manilkara hexandra*), and loss of progeny and prosperity (*Ziziphus mauritiana, Punica granatum* and *Indigofera tinctoria*). The taboos associated with inauspicious trees and planting of auspicious trees contained in *Vrikshayurveda* evoke curiosity. The text does not specifically mention why specific plant species are to be avoided. The text is also silent on the issue of planting specific trees on a particular side of the house. Perhaps, it may refer to the climate and direction of the sun and rain in relation to different rooms and their uses within the household.

Spiritual significance of trees and flowers

Shri Aurobindo and the Mother (2004) offered spiritual significance to plants. Some of their thoughts are discussed here.

Any plant has an aspiration to grow. It is not certain whether the plant makes an effort or not. Plants grow because they aspire for the light, for the sun and for the open air. There is a kind of competition among plants for sun light. If one goes into a wood, or into a park where there are many different plants, one can observe very clearly that there is a sort of competition among plants to pass each other and reach the light and open air above. It is indeed quite wonderful to see.

According to the Mother we must suppose that in the plant also, there is a force to which we can give the name of consciousness although it is not the human or animal mentality for which we have hitherto preserved the monopoly of that description. In ourselves, there is such a vital consciousness which acts in the cells of the body and the automatic vital functions so that we go through purposeful movements and obey attractions and repulsions to which our mind is a stranger. In animals this vital consciousness is an even more important factor. In plants it is intuitively evident. The seekings and shrinking of the plant, its pleasures and pains, its sleep and its wakefulness and all that strange life, whose truth an Indian scientist, Dr. Jagadish Chandra Bose has brought to light by rigid scientific methods, are all movements of consciousness, but, as far as we can see, not of mentality. There is then a sub mental, a vital consciousness which has precisely the same initial reactions as the mental, but is different in the constitution of its self-experience, even as that which is superconscient is in the constitution of its self experience different from the mental being.

There are trees, which have lived long and can be the home of a conscious being, a vital being. The plants serve as home and shelter for birds and butterflies, which are not, created by the plant itself.

Psychic power of plants and flowers

Innumerables like ideas and flowers are joyous companions. The soul of a plant or an animal is not dormant - only its means of expression are less developed than those of a human being. There is much that is psychic in the plants, much that is psychic in the animal. The plant has only the vital physical elements evolved in its form; the consciousness behind the form of the plant has not developed or organised mentally, capable of expressing itself. The animal takes a step farther; it has a vital mind and some extent of self expression, but its consciousness, its mentality and its experience are limited. All the same, animal has a soul and can respond very readily to the psychic in man.

Directly there is organic life, the vital element comes in, and it is this vital element which gives the sense of beauty to flowers. It is not perhaps individualized in the sense we understand it, but it is sense of the species and the species always try to realize it. The Mother noticed a first rudiment of the psychic presence and vibration in vegetable life, and this blossoming one calls a flower, is the first manifestation of the psychic presence. The psychic is individualized only in man, but it was there before him; but it is not the same kind as in man. It is more fluid; it manifests as force, as consciousness rather than as individuality. Take Rose, for example; its great perfection of form, colour and scent express an aspiration and a psychic giving. Opening of rose in the morning at the first touch of the sun is a magnificent self-giving in aspiration.

The Mother advised human beings to be like a flower. One must try to become like a flower; open, frank, equal, generous and kind.

A flower is open to all that surround it: nature, light, the rays of the sun, the wind, etc. It exerts a spontaneous influence on all that is around it. It radiates a joy and a beauty.

It is frank: it hides nothing of its beauty, and lets it flow frankly out of itself. What is within, what is in its depth, it lets it come out so that everyone can see it.

It is equal: It has no preference. Everyone can enjoy its beauty and its perfume, without rivalry. It is equal and the same for everybody there is no difference, or anything whatsoever.

Then generous: without reserve or restriction, how it gives the mysterious beauty and the very own perfume of nature. It sacrifices itself entirely for our pleasure, even its life it scarifies to express the beauty and the secret of the things gathered within itself.

And then, kind; It has such a tenderness, it is so sweet, so close to us, so loving. Its presence fills us with joy, it is always cheerful and happy.

Happy is he who can exchange his qualities with the real qualities of the flowers. Try to cultivate in yourself their refined qualities.

When flowers are given to a person he may develop the Divine qualities they symbolize. They can directly transmit into the psychic all that they contain, pure, unalloyed. They possess a very subtle and very deep power and influence. Each flower symbolizes an aspect, an emanation, an aspiration and a progress in the evolution of the earth.

The message of trees and their flowers

The Mother attributed a specific message to each and every flower. The message given by the flowers of the important ornamental flowering trees according to The Mother is given below.

Butea monosperma **(Flame of the forest)**

A tree with dense racemes of medium sized bright red-orange papilionaceous flowers, gives the message of *'Beginning of the Supramental realization'*, with its charming beauty, is the herald of victory.

Callistemon lanceolatus **(Bottlebrush)**

Gives the message of *'Joy of Vegetal Nature in Answer to the New Light'*, it dances with joy and laughs happily.

Cassia fistula **(Golden shower tree, Indian laburnum)**

A tree with long pendulous clusters of fragrant canary-yellow rotate flowers, gives the message of *'Imagination'*, abundant and varied, may be charming, but must not be substituted for Truth.

Cassia javanica **(Apple-blossom cassia, Java cassia)**

A tree with pendulous cluster of fragrant rose-pink flowers turning white, symbolizes *'Psychic Work'*, a work governed by harmony.

Couroupita guianensis **(Cannonball tree)**

A tree with pendulous racemes of large and very fragrant light to deep red rotate flowers with a central recurved hood, symbolizes *'Prosperity'*, stays consistently only with those who offer it to the Divine.

Delonix regia **(Flamboyant, Gul Mohur)**

A tree with a cluster of striking orange red to deep red flowers, one petal white to cream or yellow and marked orange-red, symbolizes *'Realisation'*, the goal of our efforts.

Erythrina variegata **(Indian coral tree)**

A tree with one sided dense racemes of dark red saber-shaped flowers, it gives the message *'Beginning of Realization in Matter'*, matter respondents to the Divine's influence.

Jacaranda mimosaefolia **(Green ebony)**

A tree with large panicles of lavender blue tubular flowers with flared petals, gives the message of *'Attempt at Vital Goodwill'*. An attempt is a small thing, but it can be a promise for the future.

Kigelia pinnata **(Sausage tree)**

A tree with pendulous racemes of mildly fragrant dark red open funnel-shaped flowers with thick crinkled petals gives the message of *'First Response of the Inconscient to the Divine Force'*, the first step towards transformation.

Lagerstroemia speciosa **(Pride of India)**

It gives the message of *'Intimacy with the Divine in the Psychic'* - the natural state of fully developed psychic.

Michelia champaca **(Champak)**

Spicy fragrant cream-white flower symbolizes *'Divine Smile'*. We can contemplate the smile of the Divine when we have conquered our ego.

Millingtonia hortensis **(Indian cork tree)**

A tree with loose panicles of fragrant white flowers with long corolla tube, symbolizes *'Transformation'*, the goal of creation.

Mimusops elengi **(Medlar)**

A tree with very fragrant small cream-white star-shaped flowers, symbolizes *'Patience'*, indispensable for all realization.

Saraca indica **(Ashoka, Sorrowless tree of India)**

A tree with fragrant clusters of apricot to orange-red flowers with long stamens, gives the message of *'Absence of Grief'*, the contemplation that takes you above suffering.

Thespesia populnea **(Portia tree)**

A tree with bright yellow deep cup-shaped flowers with crape like petals, changing to carmine, symbolizes '*Health*', not to be preoccupied with it but to leave it to the Divine.

3

Trees in Indigenous Landscaping Historical Background

The early inhabitants of our country were very conscious of the placement of their dwellings in beautiful surroundings. The best examples are from the West Coast, where the houses were traditionally constructed in the midst of gardens. The Arab traveller and historian, Ibn Batuta, in his itinerary in coastal Malabar recorded their picturesque beauty, as early as in the 10th century AD. The whole compound was planted with trees, which were useful and aesthetic in appeal. The scholarly elite used trees and shrubs mentioned in Kalidasa's work in their planting schemes. Through the tree and foliage, a path is formed to the main village.

The thatched houses situated in woodlands helped to beat the heat and humidity. The affluent also added a pool to further beat the heat. William Logan, a British collector and historian of the then Malabar district in his book entitled 'Malabar' gave an appreciative account of not only this architecture but also its inherent beauty and cleanliness. The tradition and written accounts go to show that there was an elite class of citizens there who cherished learning, practiced cleanliness and above all, put trees to aesthetic uses.

Forming a vista through a tree lined avenue, as a means of beautifying temple architecture was also well known in ancient days. Planting an avenue of trees along major thoroughfares for shade and direction of traffic, as a temporal duty of kings, was also in practice for a long time. *Ficus benghalensis,* the banayan, was the preferred species for this purpose.

Classical literature is a storehouse of information on indigenous landscaping. A very extensive vocabulary relating to designed landscapes, plants and planting, flowers and fruits, tree culture, natural scenery and other matters relating to aesthetics was characteristic of South Indian languages, including Sanskrit, which was the cultural language in the south.

Trees in 'sacred groves' of homesteads

Evidence of the earliest use of trees for landscaping could be found in the sacred groves of Kerala. Sacred Grove represent the major effort to recognize and conserve biodiversity (ethnic diversity) traditionally. The age old system of every village having a temple, a tank and associated sacred grove explains the ancient method of water harvesting and sharing and may be considered as the backbone of village economy. People were prohibited from felling trees and even removing a twig was considered as taboo. Sacred groves have been reported from many parts of the world where mostly tribals live and practise shifting cultivation eg. Mexico, Ghana, Nigeria, China, Syria, Turkey etc. In India they are known from Himalayas, North-East India, highlands of Bihar, Orissa, Madhya Pradesh, Andhra Pradesh, Karnataka, Tamil Nadu and Kerala. Locally they are known variously as "Sarana" (in central India), "Devrai" and "Deviahate" in Maharashtra, "Devarkadu" in Coorg, 'Orance' in Rajasthan, "Kavu" or "Nagavanam" in Kerala, "Nandavana" in Tamil Nadu, "Sidharavanam" in Karnataka, and "Kavu" in Andhra Pradesh. In some cases the tallest tree in the grove is worshipped in the belief that it is the incarnation or abode of god. Trees such as *Borassus flabellifer; Alstonia scholaris, Antiaris toxicaria, Hopea parviflora, Strychnos nux-vomica; Ficus religiosa* etc. are being worshipped in this way.

One significant aspect is that being mini forests, these 'kavus' abound in large stately trees which have grown into full size. Right from vedic period due to fetishism, trees have become objects of veneration. In vedic literature, there are references about amulets prepared with plants to ward off evil spirits and cure diseases. Similarly there are specific trees mentioned for planting and preservation. The "Sthala Vriksha" concept common in South India, is actually an ancient form of conservation of certain important species, mainly trees. Each temple is associated with a "Sthala Vriksha" which is linked to the deity. The leaves of *Aegle marmelos* (Bilva or Koovalam) are usually used for worshipping Siva and flowers of Cassia fistula to adorn Vishnu. Usually sacred groves have one or more "Sthala Vriksha". These are protected and worshipped for their sociological, medicinal and social importance. Common examples of such species are Peepal (*Ficus religiosa*) Neem (*Azadirachta indica*) Maruthu (*Terminalia paniculata*) Kanikonna (*Cassia fistula*) and Bilva (*Aegle marmelos*).

Most of the sacred groves represent the relics of once gregarious and abundant low lying evergreen forests of the Western Ghats. Only very few are reported from the foothills and the high ranges. The size of the sacred grove in Kerala varies as small as one cent to 20 or more hectares.

The common tree species found in the sacred groves are *Artocarpus hirsutus, Mesua ferrea, Vateria indica, Hopea parviflora, Alstonia scholaris, Mimusops elengi, Hydnocarpus pentandra, Holigarna arnottiana* etc. Some of the trees such as *Borassus, Alstonia scholaris, Antiaris toxicaria, Hopea parviflora, Strychnos nux-vomica, Ficus religiosa* etc are being worshipped in many sacred groves.

A peculiar aspect of the selection of trees for worship from classical times is the colour symbolism. Thus white, blue and yellow flowers are sacred to Lord Vishnu and Lord Krishna. Red and its various shades as well as orange are the favorites of Lord Siva and his consort Parvati. The golden yellow flower of *Cassia fistula* is the symbol of the New Year and prosperity.

Biodiversity of flowering trees in the Western Ghats

In many respects, Kerala represents an epitome of the Western Ghats. Located in the west coast of peninsular India between 8^0 18' and 12^0 48' north and longitudes 74^0 4' and 77^0 50' east, the Western Ghats cover only 5% of India's total geographical area but contains more than 27% (4000 species) of country's total plant species (Nayar, 1996). The Western Ghats, also known as the Sahyadris, extend about 1600 km starting in North from Tapti river and going down to Kannyakumari in the South. These ranges run parallel to West coast hardly 40 kilometres away on an average from shoreline. The Western Ghats run through the states of Maharastra, Goa, Karnataka, Tamil Nadu and Kerala. Elevation of Western Ghats is 900 - 1500 m and in some places goes beyond 2000 m. The Western Ghats form major watershed in Peninsular India and 58 major Peninsular Indian rivers originate from them. Western Ghats earned a status as one of the 25 biodiversity 'hot-spots' of our planet (Myers, 2000). Western Ghats range is taxonomically more diverse and floristically rich.

Kerala is remarkably well known for her luxuriant vegetation and rich biological diversity. Perhaps no other state in India with similar geographic and physiographic conditions has been so blessed with an immensely rich natural resource base as in Kerala. The geographical isolation of this state bordered in the west by the Arabian Sea and in the east by the Western Ghats and its peculiar, physiographic, edaphic and climatic gradients contributed significantly to the development of diverse type of ecosystems, each supporting unique assemblages of biological communities with an impressive array of species and genetic diversity.

Located in a tropical region, Kerala is blessed with equable climate which correspondingly manifests in a rich forest regime. The total forest cover of India is 40,107 sq.km. i.e., 19.47 per cent of the total geographical area. This assessment was based on visual and digital interpretation of satellite data. Accordingly, the total forest cover of Kerala is 10,336 sq.km, which accounts for 26.6 per cent of the total land area (38,864 sq.km); of this, 8421 sq.km has been classified as closed forests with more than 40% crown density. The forests in Kerala afford *in-situ* protection to some of the unique plant and animal species of the world including fascinating ornamentals. South Indian Lady's slipper orchid (*Paphiopedilum druryi*), Malabar daffodil orchid (*Ipsea malabarica*), the Silent Valley grass (*Silentvalleya nairii*), etc are a few of the well-known examples.

Western Ghat area with its proximity to the equator and sea, good rainfall from both monsoons, warm humid tropical climate and rich soil strata, wide altitudinal variations and many perennial hill streams criss-crossing the area, favoured the development of very rich tropical forest types in these ghat zones sheltering many valuable, rare, endemic ornamental, medicinal and economically important plant species. Subramanyam and Nayar (1974) estimated 3500 flowering plants in the Western Ghat area, of which about 1500 are endemic. The vegetation of the Ghat is influenced by rainfall and temperature and it shows a North to South increase in luxuriance, height and diversity of species. In general, the vegetation is more or less stunted towards the north due to lesser rainfall and prolonged dry season.

There are 490 arborescent taxa represented in the low and medium elevation evergreen forests of the Western Ghats, among which 308 (62.8%) are endemic to the region. Further, the southern part corresponding to the hills of Southern Kerala and the hills south of Palghat gap comprises 87 per cent of the total endemic tree species of the Western Ghats, constituting the main centre of endemism in Southern India.

It is believed that a few regions of the world are inherently richer in endemic species. Myers (1990) termed them as 'hot spots'. Within Kerala, we can identify a few centres with a high degree of endemism. They have some unique flora including ornamentals.

Agasthyamala

This is a towering peak of 1866 m at the tail end of the Western Ghats. Together with the adjoining mountains, it constitutes an important area of high endemism. The area has typical evergreen forests and subtropical montane forests locally called 'sholas', which occur above 1500 m. Shorter boled trees which rarely exceed 6 m height due to high velocity of wind, are densely clothed with lichens, mosses, ferns and orchids. The unique genus *Brysophyllum,* found only in Sri Lanka and South India, occurs here and is represented by the endemic species *B. tetrandrum* (Bedd) J.D. Hook. Ex Bedd. About 150 species of flowering plants are endemic to this area. The most famous among them is the South Indian Lady's slipper orchid *Paphiopedilum druryi* (Bedd) Stein. This plant is protected against illegal collection, possession, cultivation and trade.

Sabarimala

Situated in the densely forested Pathanamthitta district, Sabarimala and its environs form an ideal example of conservation due to religious regions. The famous temple of Lord Ayyappa is situated on the top of this mountain and is visited by hundreds of thousands of pilgrims every year from Southern India. The Poonkavanam, which in Malayalam means a garden of flowering plants, is maintained around the temple.

Thekkady

Thekkady and its environs form a natural home to hundreds of wild animals and plants and is the core area of the Periyar Tiger Reserve, named after the river Periyar. It is one of the oldest protected areas in India and was first noticed in 1935 as a Game Reserve. It was subsequently declared a Wildlife sanctuary in 1950 and brought under Project Tiger in 1978. All types of forests and savannah grasslands are seen here on the upper reaches. The lost orchid *Taeniophyllum scaberulum* J.D. Hook., originally reported from Kottayam, was rediscovered here. The only gymnosperm of the Southern hemisphere, i.e. *Podocarpus latifolius,* is found here. Sixty four species of grasses have also been recorded.

Eravikulam National Park

Famous for its large but highly endangered population of Nilgiri tahr (*Hemitragus hylocrius*), this is a unique high altitude montane ecosystem at tropical latitude. The uniqueness of this park lies in the presence of a large shola grassland montane ecosystem in which the shola forest represents relict tropical elements while the grassland contains extra tropical elements (Mani, 1974). Characteristic alpine elements, such as *Rhododendron, Pedicularis, Gaultheria* etc. are seen at higher reaches near Anamudi, the highest peak, south of the Himalaya (2695 m). Wight's brachycoryth orchid (*Brachycorythis wightii*) was rediscovered recently from here. The area also has vast stretches of the Nilgiri strobilanth (*Strobilanthes kunthiana*) which flowers once in 12 years.

Silent Valley National Park

At the helm of the controversy during the early 1980s Silent Valley was known to be synonymous with the lion tailed macaque (*Macaca silenus*) which was once abundant. When the controversy began, we had scanty information about the flora and fauna of Silent Valley. The grass genus *Silentvalleya* and its associate orchid *Ipsea malabarica* are known among the significant species making up the rich plant wealth of Silent Valley.

Wayanad Wildlife Sanctuary

The second largest Protected Area in the State after Periyar Tiger Reserve, Wayanad Wildlife Sanctuary was established in 1973. It spreads over 344 sq. km and is contiguous with the Nagerhol Wildlife Sanctuary in Karnataka and Mudumalai Wildlife Sanctuary in Tamil Nadu. It also forms part of the Nilgiri Biosphere Reserve. Extensive patches of bamboo provide excellent habitats of elephants. Jerdon's orchid is found exclusively in Wayanad. This sanctuary has the largest tribal population in Kerala.

Indigenous trees of Western Ghats

A brief description of trees indigenous to Western Ghats which are already used for ornamental purpose are given below.

***Albizia lebbek* (Vaka), Family: Fabaceae**

Handsome, deciduous, fast growing and spreading tree, vaka is also used as a shade tree. This tree was also associated with the sacred groves (sarpakavu) of Kerala. The tree grows upto a height of 20-25 m. The tree is somewhat resistant to drought. It is propagated from seeds and is used as a shade tree in tea estates.

Alstonia scholaris (Ezhilam pala), Family: Apocynaceae

The tree which is commonly known as 'Devil's tree' was commonly associated with the sacred groves (sarpakavu) of Kerala. It is tall (16-30 m), leaves are elegant, whorled and branches provide plenty of shade. The flowers are long, slender; drumstick like fruits cover the whole tree during summer. Flowers are white or greenish white and strong smelling during night. It grows well without care and is propagated from seeds. The wood is used in packing cases. Bark is used in medicine.

Anthocephalus cadamba (Kadamba), Family: Rubiaceae

This tree is associated with Lord Krishna. Vast forests of *kadamba* trees were in existence in the region of Vrindaban, between Madura and Bharatpur, remnants of which can still be seen in this region. It is a large graceful deciduous tree with a straight stem about 30 m high and with a girth of 2 m. The spreading crown bears drooping branches and yellow ball like flowers in globose heads amongst shiny broad leaves. The leaves are long, elliptic oblong, or ovate, acute or acuminate, shining and glaborous above and pubescent underneath. Flower heads are solitary, terminal and globose with short penduncle. The calyx lobes are oblong, pale greenish or cream coloured and persistent. The corolla is glabrous and orange coloured. The flowers appear from June to August in the rainy season. The false fruits (pseudocarps) are large, fleshy and orange coloured. The fruits ripen towards the end of rains. The fruits are acidic but pleasantly flavoured and cherished by various animals, including flying foxes which also help to disseminate its minute seeds.

The tree is admired for the golden balls of flowers and for their delicate scent. When planted in large groves, the flowers create a beautiful effect. It is chiefly used for packing cases, tea boxes, beams and rafters and is also suitable for dugouts, canoes, carving and turnery. It is also considered suitable for cheap paper making. Foliage is lopped for fodder. The tree is commonly grown in moist deciduous forests and thrives the best in well - drained soil. It grows rapidly in the first 6-8 years and attains the maximum size in about 20 years.

Bombax malabarica **(Simul or Red silk cotton tree), Family: Bombacaceae**

Often a very large deciduous tree, reaching 40 m. or more in height, with trunk and branches covered with warty spines, buttressed. Leaves compound, digitately arranged, leaflets 3 to 7 in number, 5 to 12 cm. long, wide, lanceolate or oval in shape, tip pointed and base acute. Flowers large, crowded at the end of the branches appearing when the tree is leafless. Petals five, elliptic oblong, recurved, fleshy. Calyx thick, cup shaped, usually 3 lobed. Stamens arranged in five bundles of 9 to 12 and a central bundle of stamens. Fruit is a capsule filled with long fibres of silky cotton and small round seeds.

The tree yields a dark brown astringent gum used in medicine. The inner bark of the tree yields a good fibre suitable for cordage. The wood is used in the manufacture of matches. The most important product of the tree is the floss used for upholstery and life belts.

Butea frondosa **Koenig (Palas or Flame of the forest), Family: Fabaceae**

A small to medium sized, 5-10 m tall, untidy deciduous tree with crooked trunk and branches. The leaves are trifoliate, each leaflet being large, rounded, leathery, and rough. The flowers are large and crowded on leafless branches. The flowers are bright flaming scarlet orange with dark olive green to black calyces, with the shape resembling a parrot's beak. The flowering season is between April and May. It flowers a little earlier in and around Bangalore. The tree is suitable for planting in small groups in a garden or a park. Propagation is from seeds which should be sown when fresh. There is a rare variety, *B. monosperma* var. *lutea,* with yellow flowers. The tree withstands frost and drought very well, does well on saline soils. The commercial shellac is produced on this tree. The bark is used for tanning. When several leaves are stitched together it serves as a plate. The flowers yield a dye.

Cassia fistula **Linn. (The Indian laburnum) Family: Fabaceae**

Cassia fistula is the State flower of Kerala. It is small to medium sized, open branched, deciduous and hardy tree with smooth greenish grey bark on young trees, turning to brown and rough in older trees.

Leaves are large and coarsely pinnate. The fresh leaves are often of a rich copper colour with a soft downy undersurface, and remain pendulous until fully grown. Leaves fall in March - April and the first half of May. It produces long golden yellow pendent flowers which appear in April - May. It is nearly leafless at the time of flowering, the first splashes of gold appearing as the last of the old leaves are shedding, and the flowering continues till the new leaves develop. Fruit pods are long and rounded and are dark brown. The wood is hard and is used in making crafts and agricultural implements. Bark is used for tanning. The dark brown sweetish pulp is laxative and is used medicinally. It is suitable for parks, gardens and avenues.

Dillenia pentagyna (Dillenia), Family: Dilleniaceae

Dillenia, is named in honour of J.J. Dillenius (1684-1747), a noted botanist, *Pentagyna* in allusion to the flower having five styles. It is a moderate sized deciduous tree, 22-25 m. high. Leaves are crowded at the end of the branches, smooth and shining, nerves prominent ending in sharp points, leaf stalk reaching, embracing the branch. Flowers arising in clusters from small tubercules on the old wood of the branches, petals with cranked margins, stamens numerous, fruit the size of a ping pong ball, succulent. Propagated by seeds, it requires plenty of light. It is sensitive to frost, but is able to withstand forest fire. The large leaves are used to form a substratum for roof thatch. The fleshy sepals are eaten by the hill people. The tree flowers during the early part of the year at the time when it is quite leafless.

Erythrina indica Lam. (Indian coral tree or Murikku), Family: Fabaceae

Erythrina is derived from a Greek word meaning red, in allusion to the colour of the flowers and *indica* means pertaining to India. It is a fast growing, densely branched small tree. The erect stem has black spines. The trifoliate leaves give the plant an ornamental look. The tree flowers in a defoliated condition during February to March and even upto April. The scarlet red flowers appear in racemes in a cluster. The tree is planted for the beauty of flowers. It can be planted in groups in parks and gardens or along avenues in towns and cities. The tree being small, it is possible to plant them in multiple rows.

***Erythropsis colorta* Burk. (Scarlet sterculia), Family: Sterculiaceae**

Erythropsis in Greek means red colouring of the flowers, *Colorata* means coloured in reference to the bright colouring of the flower. A moderate sized deciduous tree, leaves large, broader than long, variable in shape, commonly 3 lobed, lobes triangular, acuminate at the tips, base deeply cordate, stalk long, nerves 3 to 5 from the top of the petiole to the tips of the lobes. Flower about 3 cm long, many together appearing at the end of the branches when the tree is leafless. Calyx funnel shaped with small triangular lobes, covered with a scarlet red 'dust'. Fruit composed of 3 to 5 papery follicles, leaf like, elliptic, veined, with two seeds attached to the margin of the follicle. The tree is propagated by seeds. Seeds should be planted soon after they mature. The bark yields an inferior fibre, strong but coarse, occasionally used for ropes. The leaves are used as fodder.

***Ficus benghalensis* Lin. (Banyan tree), Family: Moraceae**

A very large, spreading, deciduous tree, often reaching 25 m. or more, sending down numerous aerials from its branches to the ground which in time support the branches. Leaves egg shaped, rounded or elliptical, tip blunt, base rounded or somewhat heart shaped, 3 to 7 nerves arising from the base. Flowers minute within the 'fig' (or receptacle), fruit a fig. It is widely used as a shade tree. The juice of the Banyan yields an inferior rubber and is used in medicine being applied externally for pains and bruises and for rheumatism. An infusion of the bark is regarded as a powerful tonic in the treatment of diabetes. The leaves are heated and used as a poultice. The fruit is eaten as a famine food. The twigs and leaves make good fodder. The wood is of little value, but as it is durable under water, it is used for well curbs.

***Holarrhena antidysenterica* Wall (Kodaka pala), Family: Apocynaceae**

A small tree commonly grown for afforestation. Leaves, bark and seeds are used to cure dysentery and diarrhoea. Although slow in action with emetine, bark is less toxic. Wood is used for shoe - heels, pen holders, combs, engraving, toys and model making. It is a shade tree in gardens.

***Hymenodictyon obovatum* (Kadva-sirid), Family: Rubiaceae**

The name *Hymenodictyon* is in reference to net veined character of the bract and *Obovatum* in allusion to the ovate shape of the leaves. It is a large deciduous tree reaching 15 m in height. Leaves elliptic or obovate, abruptly acumiante, glabrous above, more or less pubescent beneath, 6-8 pairs of main nerves, long petioled. Flowers small, inconspicuous, creamy sessile, densely crowded along the rachis, inflorescence panicled. Stamens 5, ovary 2 celled; stigma fusiform, capsule many seeded. The bark yields a bitter extract used in Indian medicine as a febrifuge. The wood is employed for making agricultural instruments. A characteristic of the tree is the large leaf like bract below the inflorescence which is creamy white or pale green when fresh, but dries hard and papery and remains erect and rattles in the breeze long after the leaves have fallen.

***Lagerstroemia flos-reginae* (Pride of India), Family: Lythraceae**

This tree is 8-10 m in height, but in its natural habitat, it sometimes grows upto 20 m. The leaves are large (15 cm x 8 cm) and handsome. During February - March the leaves turn somewhat coppery or yellowish in colour. The showy, purplish, purple or mauve flowers are borne in about 30 cm long panicles at the end of all branches during April - May and July - August. Propagation is from seeds. The seedlings are transplanted after one year. The growth is slow during the first couple of years, but afterwards it grows rapidly. The flowering starts in about five years. The plant grows best in deep alluvial loam. Recommended for growing along avenues, and in parks and gardens.

***Madhuka latifolia* (Syn. *Bassia latifolia*) (The Mahua or Indian butter tree), Family: Sapotaceae**

It is a large deciduous tree with wrinkled bark which may be of grey or brown shade. Leaves are leathery and clustered near the ends of branches. Young leaves are coppery red and beautiful in February - March. It is an excellent tree for avenues where shade is required. Flowers of the tree are cream coloured. They appear from February to April when the tree is almost leafless. The tree blooms at night and flowers fall to the ground at dawn. The fruit is a green juicy berry, ripening in June - July and is edible. Mahua is a forest tree, which is also cultivated for its nutritious flowers which are the staple food of the forest dwellers of central India. Its virtues are celebrated in numerous folk songs from central and eastern India. Seeds on crushing, yield oil which is used as a substitute for *ghee,* hence the name Butter Tree. Mahua oil is also used for illumination. Mahua cake is used as manure. It is also used for expelling earth worms from tennis courts. The bark is used as a dye. Its milky sap from cut flowers, stems and branches is used as a cure for rheumatism.

Michelia champaca (Golden champa), Family: Mangnoliaceae

A medium sized tree, 10-13 m tall, having long leaves with wavy margins slightly resembling mango leaves. The 5-7 cm long flowers are golden coloured, highly scented and silky in appearance. The flowering season is between April and May and September and October. The white flowered species mentioned above and a creamy yellow variety are propagated by grafting on the golden flowered variety which produces seeds in abundance. The seedlings take about 7-8 years to come to flower. When a plant tends to bear abundance of fruits, it should be removed, otherwise the plant may be exhausted and may not bear flowers properly for the next couple of years. The tree likes a humid atmosphere and grows upto medium altitudes, suitable for planting along avenues and in parks and compounds.

Mimusops elengi Linn. (Elengi), Family: Sapotaceae

A large evergreen tree with a dense crown of dark green, shining foliage, 20 m high. Leaves elliptic, tip shortly acuminate, base acute or rounded. Flowers star shaped, fragrant usually turning downwards. Stamens 8, opposite to the inner circle of petals. Fruit egg shaped, turning orange when ripe. The tree flowers sporadically from July to December, but the general period appears to be around November to December. The tree yields the pogoda gum of Madras. The bark is used either by itself or in combination with other barks as a dye. The flower contains a volatile oil from which a sweet scented water is distilled. An oil is also obtained from the seeds. A mouth wash and a gargle are prepared from the flowers and fruits. The fruit, when ripe, is eaten. It contains a small quantity of sweet pulp. A preserve is also prepared from the fruit. The oil from the seeds is used for cooking. The wood is good, and is used in building and for carts and cabinet work. It is planted in large garden.

***Pongamia glabra* Vent. (Karanj), Family: Fabaceae**

A moderate sized deciduous tree reaching 15-20 m high commonly planted as a road side tree. Leaves compound, imparipinnate, leaflets opposite with an odd one, somewhat egg shaped or elliptic, acute or sometimes acuminate at the tip, base acute or rounded. Flowers are produced in short racemes in the axils of the leaves, which are shorter than the leaves. Calyx small, cup shaped. Fruit is a pod, 1 to 2 seeded. It is a very hardy tree and can adapt itself to varying conditions. The flowers and new leaves generally appear together. The leaf shedding is very irregular in the species depending possibly on situation and age. When the trees are in flower, the ground below them is usually carpeted with blooms. The fruit remains on the tree till the next lot of new leaves appear.

***Rhododendron arboreum* Smith. (Rhododendron), Family: Ericaceae**

It is a large shrub or tree, attaining a height of about 15 m. The leaves are 10-12 cm long, oblong to lanceolate, acute, rugose above, distinctly veined and whitish beneath. The flowers are in dense clusters and are of red colour. The pedicles are short with minute calyx and campanulate corolla. It flowers in March to May, when hill-sides are a scarlet gleam. Flowers are acid in taste and are used for chutney making by the hill people.

***Saraca indica* Linn. (Asoka), Family: Fabaceae**

A small sized, 7-10 m tall handsome, evergreen tree, with branches spreading in all directions and forming a beautiful crown. The leaflets are 3-7 pairs, opposite, lanceolate and 12-15 cm long. The young leaflets come out in drooping bunches, are greyish white to pinkish in colour, and remain folded. The rich orange, slightly fragrant flowers are borne in roundish panicles on the branches and stems. The flowering season is the summer. All the *ashokas* are easily propagated from seeds. *S. indica* is suitable for planting as a specimen in lawns or in groves in a park.

Terminalia arjuna Bedd. (Arjuna), Family: Combretaceae

Terminalia refers to flower appearing at the ends of the branches, *Arjuna* is derived from one of the vernacular names for the tree. A large deciduous tree with thick, often buttressed trunk and smooth whitish bark. Leaves mostly sub opposite, oblong or ovate oblong, blunt or slightly pointed at the tip, often crenulate, base rounded or cordate or occasionally unequally sided; petiole 5-6 cm with two prominent glands at the top, under the blade. Flowers sessile, in short axillary spikes or in small panicles. Fruit with 5 leathery wings and tan coloured when dry. Propagated by seed. Although not a garden tree, it is frequently planted as an avenue tree.

The wood is used in the construction of carts, agricultural implements and boat building. It is also used for support in mines. It yields a brown transparent gum used as a drug in North India. The bark is employed in medicine as a tonic and also for dyeing and tanning.

Thespesia populnea Soland ex Correa (Syn. Hibiscus populneus Linn) (The bhendi tree), Family: Malvaceae

Thespesia is derived from the Greek, the *spesios* meaning divine, so named because of its being often planted near churches or temples and *populnea* refers to its poplar like leaves. It is a large evergreen tree (10 to 15 m) with crooked stem and poplar like glossy, dark green leaves. Flowers large, yellow and cinnamom red, bell shaped, resembling those of the vegetable plant *bhendi* or cotton plant. They appear singly, off and on, throughout the year. Peak flowering is in cold weather. The fruit is a capsule, somewhat rounded and flattened at the top. Pods are filled with white papery seeds. It is a shady road side tree and is also suitable for medium sized gardens. Its bark yields a strong fibre. Its wood is tough, with fine grains, and is used for making gun stocks and cart wheels. Leaves are used for wrapping food. Seeds yield a thick oil. Its fruit juice is used for external application to cure cutaneous diseases. Root is taken as a tonic in Malaya, the deep red heartwood is said to be used as a remedy in heart attack. It is easily raised from seeds or from cuttings. Growth is rapid. It prefers a light porous soil and is extremely salt resistant, suitable for saline soils. It is one of the rare trees which flower all the year round, and is also shady. It deserves greater popularity. On account of its hardy nature, it can grow in all the states in India, except the hills.

4

Functional and Aesthetic Values of Trees

Trees alone on the earth give happiness both here and hereafter. Since they "save" from abject poverty they are named "taravah" (the saviors). Therefore, the trees should be carefully nurtured as through their shade, flowers and fruits they help immensely in the pursuit of dharma, artha, and kama (Vrikshayurveda).

The tree that lives for 50 years

- Generates Rs. 5.3 lakhs worth of oxygen,
- Recycles Rs. 6.4 lakhs worth of fertility,
- Facilitates Rs. 6.4 lakhs worth of soil erosion control,
- Makes Rs. 10.5 lakhs worth of clean air and
- Provides Rs. 5.3 lakhs worth of shelter for birds and animals.
- Besides, it provides flowers, fruits and timber
- So when one tree falls or is felled,

We lose something worth more than Rs. 32 lakhs and our *fragile nature*

Courtesy: Wasteland News

It has long been thought that landscape plants are used only for beauty without regard to their functional value, and the term 'ornamental plant' is indicative of this limited concept. But in modern landscaping, plants can be used functionally to improve the quality of an environment.

The quality of life for many of the Indians is dictated by the quality of their city or town. As a result of burgeoning population and increasing trends towards urbanization, heavy pressure is put on the civic amenities available in cities. The hectic pace of life in cities, keeps people pre-occupied with jobs, education, crime and similar other urban issues that importance of environment in urban life gets overlooked. As a result, most cities are getting far more polluted, hotter and less pleasant than they need to be. Improving these conditions can make life better for every one. Planting of ornamental trees decorates urban landscape and adds to the bio-aesthetic value of town planning. It has a definite curative effect on human mind engulfed by busy and congested city life.

Functional values of trees

Since the time he appeared on this planet, man depended on plants for many of his needs including primary requirements of food to keep himself active and alive. It is beyond doubt that the trees top the list of plants useful to mankind. Trees provide human beings with fruits and nuts for his food, timber, drugs, paper, rubber, cork and other materials for industry, ornamentals for the decoration of the gardens and hundreds of other products for meeting his daily needs.

Each letter of the word "Tree" has a logical meaning, especially with regard to the urban environment (Nowak, 1995).

T - Temperature and microclimate modification
R - Removal of air pollutants
E - Erosion control
E - Energy conservation

Growing trees is the cheapest and one of the most effective methods of preservation and improvement of our physical environment. At present our densely populated urban areas are treeless and bare. Trees function as shelter from scorching heat and glare of summers as well as sleet and blast of winters thus serving as natural air conditioners. Trees also purify dirty air, reduce dust, noise; screen undesirable views and develop a healthy and beautiful environment. Thus the uses of trees in human settlements in the form of groves, in rows and other settings enhance the beauty and functional value of landscapes. They function as living building materials to create and reinforce urban space. The following are some of the functional and aesthetic values of trees in landscapes.

Visual control

Trees can be very effectively used in landscape to achieve visual control. Judicious planting of trees can reduce glare, both from direct and reflected light, provide privacy in the landscape, screen objectionable views and enframe views.

Glare and reflection control

Trees can be effectively used to reduce glare in the landscape. Light from sources like solar radiation, headlights of vehicles, streetlights, flood lights and neon signs directly causes primary glare. On the contrary, secondary glare is the reflected light that has been bounced off from a surface. Secondary glare is a general problem in urban areas where extensive paving and reflective materials are present.

Planting trees in the median strip of highways reduces glare from oncoming headlights of vehicles. Dense planting of this kind is particularly important where curves in the road are such that headlight beams are aimed directly at the wind shields of oncoming vehicles. Likewise, plants can be used to reduce glare from the sun or from stationary artificial light sources by locating plants of the proper height and density strategically between the source of light and the area upon which the light shines, whether it is a bedroom or outside living area.

Reflected light can be reduced by intercepting it either before it touches a reflective surface or between the reflective surface and the viewer. Trees located over reflective paving serve to filter the sunlight before it reaches the surface. But in case of reflective surface being water, it is intercepted by planting trees before it reaches the viewer.

Screening for privacy

Privacy control is the technique of enriching a well-defined place with trees and shrubs of such a height that views into and from the space are prevented. The purpose of privacy control is to isolate the space from its surroundings. Privacy control is often a desirable design objective for an intimate setting area or the terrace of a residence. A dense planting of vegetation over 2 m tall usually provides the greatest sense of privacy, while planting at chest - height provides partial privacy. Every one wants a fast growing, long lived, evergreen and maintenance-free green plant to create a feeling of privacy and security.

Screening objectionable views

One of the frequent uses of trees in landscape is to screen objectionable views. Trees as vertical barriers can control views so that desirable points in the landscape are observed while ugly points are blocked. Long lived and fast growing evergreen trees are most successful for establishing a permanent wall of vegetation that prevents views throughout the year. Further, trees planted effectively will soften the harsh lines of the house or landscape.

Enframement

Trees can be used to enframe a view towards the distant landscape. This may be done by planting trees so that they reveal an outstanding view gradually as one moves through the planting. Trees planted in front corners of buildings or on two sides of the buildings will effectively enframe the building.

Physical barriers

Trees planted in a line and sometimes pruned to a particular height comprises their use as an outdoor wall. It can be used to provide protection from animals and intruders in a landscape. Properly positioned, it can be used to control pedestrian traffic under the landscape.

Climate modification

One of the basic functional uses of plants in the landscape is to modify or control the microclimate. The four climatic factors that affect human comfort are solar radiation, air temperature, air movement or wind and humidity. By the proper uses of trees, a human comfort zone can be created that may ameliorate the climate sufficiently to be comfortable.

Solar radiation and temperature control

Trees ameliorate air temperature in urban environment by controlling solar radiation. Tree leaves intercept, reflect, absorb and transmit solar radiation. Their effectiveness depends on, for example, the density of species foliage, leaf shape and branching patterns. Deciduous trees are very instrumental in heat control in urban settings in temperate regions. During summer, they intercept solar radiation and lower temperature. In the winter, the loss of their leaves results in the pleasant warming effects of increased solar radiation.

Trees can be considered as nature's air conditioners. Trees reduce the summer air temperature through evapo-transpiration. A single tree may transpire 88 gallons of water per day which is compared to five air conditioners with a capacity of 2500 k cal/hr, running 20 hours a day.

Wind protection and air movement

Use of plants for windbreaks is well-known as an effective method of controlling wind. It depends on height, density, shape and width of plantings, but their height is the most important consideration. Windbreaks perpendicular to prevailing winds may reduce wind two to five times, the height of the tallest trees in front of the barrier and for distances of 30-40 times on the leeward side. Maximum wind reduction occurs in leeward distance of 10 to 20 times the height of the tallest trees. Taller the tree, greater the protected distance. As trees increase in height, they generally become more open and allow increased airflow in the lower portions. Here more rows of trees in combination with shrubs are usually required to provide adequate protection.

Density and shape of the windbreak are also important considerations in reducing wind velocity. A moderately dense barrier is the most effective and actually reduces wind speed over a greater distance than does a windbreak of higher density plantings.

Control of precipitation and humidity

Trees provide temporary shelter from rain during short showers. A considerable amount of precipitation collects on the foliage of trees before it passes through foliage. This can increase infiltration and distance run off and soil erosion. Trees also reduce soil moisture evaporation.

The effect of plants on atmospheric humidity is of interest in climate modification. Plants contain large quantities of water that are added to the air through transpiration. The more foliage, greater is the amount of water added to the air.

Ecological value

Noise control

Noise, as 'invisible pollution' has become a problem of increasing concern in man's outdoor environment owing to urbanization and technology. Ability of plants to control noise is determined by the intensity, frequency and direction of noise as well as the location, height, width and density of planting. Places in which trees can be effectively used to reduce noise are along highways, around industrial areas, residential areas and in parks and recreation areas of cities. Size and density of planting are critical in noise control, hedge and other narrow planting being relatively ineffective. Plantings should be atleast 6 m wide. The height is also critical, since sound will travel over the top of a barrier that is low. Proper selection and spacing of plants and proper land grading can contribute significantly to the alleviation of noise.

Pollution control

Trees play a potential role in removing atmospheric contaminants. They act as 'natural filters' which clean the atmosphere. Trees aid in removal of airborne particles such as sand, dust, fly ash, pollen and smoke. Leaves, branches, stems and their associated surface structures tend to trap particles that are later washed off by precipitation. Trees also aid in the removal of air borne particulate matter by air washing. Transpiration increases humidity, thus aiding in the settling out of airborne particles. The results of these processes can be readily observed on trees adjacent to factories or long gravel roads.

Erosion control

The thick network of roots bind the soil and keep it in place. Huge amount of water vapour is given out through transpiration that cools the atmosphere and adds humidity to make it comfortable. This water vapour also helps in the formation of clouds that bring rain.

Rain water pouring down on a clean piece of land moves unrestricted downward to be washed into rivers causing soil erosion on its way. The sifted soil results in the silting of water channels and rivers giving rise to flood that brings destruction and misery to man. On the other hand, when rain falls on a land with trees planted, its free flow is restricted. A good part of it percolates down into the soil and conserves there to feed the vegetation of the immediate vicinity during the drought period. Another part goes further down, when tapped, meet the water requirement of man and the crop grown even hundreds of miles away.

Aesthetic value

Trees provide their own inherent beauty in all settings. They are aesthetic elements in our surroundings. They can be beautiful simply because of lines, forms, colour and textures they project. Trees enframe views, soften architectural lines, enhance and complement architectural elements, unify divergent elements and introduce naturality to otherwise stark settings. They also produce unique pattern through reflection from glass and water surfaces and can produce beautiful shadow patterns. Trees are also dynamic, giving different appearances in the changing seasons and throughout their span of life. They also provide movement and pleasant sound - the rustling of leaves and the whistle of winds throughout a canopy. We are attracted to trees because of many of these characteristics.

The aesthetic value of trees can be projected in landscapes by planting as specimen trees or arranging in lines or groups.

Specimen or accent plant

Using plants as specimens is a good way to emphasize a character. An unusually beautiful tree if used sparingly, as accent plant creates interests and contrasts. Too my specimens destroy this effect.

Line planting

Straight or curved lines of tree plantings can shade parking or play areas apart from serving as windbreaks. Select hardy, well-adapted plants to form a line since the visual effect can be spoiled if one or a few of the plants are damaged.

Group planting

Grouping is a relatively natural way to the plants. Place several trees, fairly close together for a more massive effect. Since plants are often grouped naturally, this is a good method to relate your landscape to a natural environment.

Mass planting

Mass planting is really an extension of the group. In mass planting individual plants tend to lose identity. Mass plantings are especially useful in relating large buildings to the site.

Trees to attract butterflies and birds

Gods, demons, gandharvas, ghosts, snakes, rakshasas, beasts, birds, and human beings always take resort to the trees (Vrikshayurveda).

Attracting butterflies, humming birds and other wild birds often go hand-in-hand with landscaping, as plant-lovers usually appreciate wildlife, too. Plant trees useful for attracting birds, and enjoy the colour and song they bring to your yard. Like bird watching, watching butterflies is a natural extension to landscaping.

Bird watching and landscaping complement each other beautifully. The trick is knowing which blooming specimens are the most useful for attracting wild birds. Many decorative berries, while inedible for humans, are a boon to birds. Some of these berries will be snapped up quickly by them, while others, serve as emergency food during hard times.

Bee keeping

Bee keeping can go side by side with growing flowering trees. While the beneficial effects of honey are well understood, the mysterious power of pollen and bee bread and their nutritional value are less popular among common people.

Uses of pollen today

Pollen grains are small, male reproduction units (gametophytes) formed in the anthers of higher flowering plants. Pollen is transferred onto the stigma of a flower (a process called pollination) by either wind, water or various animals (mostly insects), among which bees (almost 30,000 different species) are the most important ones.

Innumerable stories and even more rumours exist about the mysterious powers of pollen and its nutritional value. Pollen is frequently called the "only perfectly complete food". High performance athletes are quoted as eating pollen, suggesting their performance is due to this "miracle food", just as the "busy bee" represents a role model for an active and productive member of society.

Pollen as medicine and as food

In order to desensitize allergic patients, pollen is usually collected directly from the plants, to allow proper identification and purity. A pollen extract is then injected subcutaneously. Desensitization through ingestion of pollen is claimed, but has not received any scientific confirmation. For treatment of various prostate problems, pollen is usually prescribed in its dry pellet form as collected by the bees. Pollen from different countries or regions seems to work equally well. However, pollen is not officially recognized as a medicinal drug. Major use of pollen today is as a food or, more correctly, as a food supplement.

In cosmetics

Pollen has only recently been included in some cosmetic preparations with claim of rejuvenating and nourishing effects for the skin.

For pollution monitoring

Since the 1980's, experiments showed that pollen collected by honeybees reflected environmental pollution levels when examined for metals, heavy metals and radioactivity. Contaminants can be quantified and sampling may be cheaper than most standard methods currently in use. Attempts are made to use pollen-collecting honeybees for identification of potential mining areas.

Beebread

The pollen stored by honeybees undergoes lactic acid fermentation and is thus preserved. Its characteristic sour taste together with brood and honey is a delicacy consumed directly during harvesting. This final storage product is called beebread. The nutritional value of beebread is much higher in places where limited food variety or quantity creates nutrient imbalances. It is particularly children who might benefit the most from regular pollen supplements in their diets.

Table 1. The average composition of dried pollen

Composition	Bee-collected		Hand-collected
	%[a]	%[b]	%[b]
Water (air-dried-pollen)	7	11	10
Crude protein	20	21	20
Ash	3	3	4
Ether extracts (crude fat)	5	5	5
Carbohydrate			
Reducing sugars	36	26	3
Non-reducing sugars	1	3	8
Starch	-	3	8
Undetermined	28	29	43

Table 2. Minor components of bee collected pollen

Flavonoids	At least 8 (flavonoid pattern is characteristic for each pollen type)
Carotenoids	At least 11
Vitamins	C, E, B complex (including, niacin, biotin, pantothenic acid, riboflavin (B_2), and pyridoxine (B_6).
Minerals	Principal minerals: K, Na, Ca, Mg, P, S. Trace elements: A1, B, C1, Cu, I, Fe, Mn, Ni, Si, Ti and Zn
Terpenes	
Free animo acids	All
Nucleic acids and nucleosides	DNA, RNA and others
Enzymes	More than 100
Growth regulators	Auxins, brassins, gibberellines, kinins and growth inhibitors

Selection of trees

Selection of trees depends on many factors which should be kept in mind when selection is being done. The purpose of growing or the functional use of the tree is a very important consideration. We have to consider whether it is used as a specimen tree, shade tree, flowering tree, avenue tree, for screening purpose or for checking air pollution. Space availability, climatic factors, soil factors, availability and cost factors are other considerations. Suitable species and varieties should be selected accordingly.

Trees are available and usable as native, exotic, or naturalized materials. *Native* plants are those which evolved through nature within a certain location. *Exotic* plants are those that have been introduced to an area by individuals, not nature and adapted very well to the introduced country. Exotic plants have certain advantages over many of the native plants. They often transplant more easily than the natives. Over 60% of the trees and shrubs used in American landscaping are exotic in origin. *Naturalized* plants are those which were brought into an area as exotic, but have adapted so well that they have *escaped cultivation.* This means that they now occur commonly both in and out of planned landscape settings.

Native plants are rightly gaining prominent place in the landscape. Although we saw an increase in planting and preserving native plant species, we are still a long way from maximizing their potential. Often the native plants are more resistant to drought, insects and diseases. If species are selected that are native to the locality, there is an additional bonus of visually relating the new landscape to the natural environment.

Criteria for selection

To function correctly, trees must be selected with careful thought, not merely because the designer happens to like their appearance. Selection of a tree is determined by a combination of several factors:

- The height of the tree
- How low to the ground it branches
- The density of its foliage
- Whether it is deciduous or evergreen
- Seasonal colour, foliage texture, and whether it bears flowers and/ or fruit.
- Hardiness
- Ease of transplanting
- Reaction to insects or diseases

One important reason for selecting a particular tree is it hardiness. The *hardiness rating* determines whether or not the plant will survive the winter in the location desired in cold countries. Some plants will not survive if the winter is too warm, while other may survive in colder areas than expected because they are planted in sheltered locations. These small and unusual planting areas are known as microclimates.

The shape of silhouette of the tree may also determine its suitability for use in the landscape. It may be wide oval, vase shaped, pyramidal, round, columnar or weeping. Based on the shape of silhouette, trees may be selected for different landscaping uses.

Closely associated with the silhouette of a tree is its branching habit or the way in which the branches tend to grow. Some trees have low, sweeping branches, such as in willows. Such trees are difficult to walk beneath, picnic or play. When human activity is expected to occur beneath trees, a high-branching species should be selected. For example, small and high branching trees (8-9 m) make excellent patio trees. They provide shade while keeping the ceiling effect low and intimate.

Size of the tree is another important measure of its suitability for the landscape. A tree that grows to 30 m tall would dwarf a small, one-storey house. However, that same tree could be attractive if planted near a much taller building. To assure that the tree selected is in the correct proportion to the rest of the landscape and any surrounding structures, the design principle of scale and proportion must be applied.

The texture and density of the plant's foliage should be considered when analyzing the qualities of tree. Texture may be determined by touch, sight, or both. Visually, texture is a combination of the size of the leaves and the shadow effect of sunlight falling upon the leaves. Foliage density determines how much sunlight or wind actually passes through the foliage. As a general rule, large trees with coarse texture (large leaves, heavy shade effect) look the best when placed away from small buildings. Trees with finer textures (small leaves, less shade effect) are suitable for use nearer to buildings.

When selecting a tree for a landscape, the foliage density should be considered in relation to the function of the tree. For example, a deciduous tree placed off the southwest corner of a house provides cooling shade on hot summer days. Later in the winter, it drops its leaves and allows the sunlight to warm the house, thereby reducing fuel consumption.

Seasonal colour is also a consideration in tree selection. A few trees give bright colour during several season in the same year from the flowers, leaves, or fruit. Colour can also be present on bark of the tree. The white, peeling bark of the white birch is a very appealing characteristic. This tree contrasts effectively with evergreens planted behind or beneath. Likewise, the shiny red bark of cherry trees and the silvery gray bark of the red and silver maple trees are important reasons for selecting these trees for use in the landscape in western countries.

Design principles in tree planting

While planting trees in the landscape, one should have to consider the five basic principles of landscape design, viz., simplicity, proportion, balance, focalization and rhythm and line.

Simplicity

The first principle of design is simplicity. Simplicity may be incorporated in several ways. Repetition of the same species of plant, construction material, or colour is one of the easiest ways to keep a landscape simple. It is not necessary, as some individuals believe, to use a great number of plants. Massing plants is another way to create simplicity. Massing can give the landscape a sense of unity not other wise possible, since the individual plants do not have to compete with each other for attention.

Balance

The second principle of design is balance. There are two types of balance useful in designing with plants; symmetric and asymmetric. With symmetric balance, one side of the landscape is planted and built exactly the same as the side opposite to it. It sometimes creates a formal appearance which may be undesirable. For this reason, asymmetric balance is more commonly used. Asymmetric balance creates the same amount of interest on both sides of the outdoor room, but does not create an exact duplication.

Proportion

Third among the principles of design is proportion. Proportion is concerned with the size relationship of the features in the landscape. A large tree is out of proportion with small house, but is in proper proportion with a large and tall house. A tree planted in proportion to its surroundings seems appropriate and comfortable.

Focalization

The principle of focalization is applied to every landscape in one or two selected spots. It is based upon the knowledge that when the human eye views a scene, it is attracted immediately to one feature, then gradually begins to take in the adjacent items. The feature which first attracts the eye is known as a focal point. It may draw attention through its shape, colour, size, texture, sound, or motion. Trees with attractive flowers function as focal points.

Rhythm and line

The final design principle is rhythm and line. This principle is used to create a sense of movement for the viewer's eye. Gently rolling tree silhouettes and stepped plant arrangements are methods by which this sense of motion is created within a landscape.

Table 3. Tree silhouettes and their landscaping uses (Ingels 1978)

Type	Character	Landscaping uses
Wide oval as in *Ficus religiosa*	• spread more than its height • often a small tree • horizontal branching pattern • branches low to the ground	• focal point plant • for framing and screening • Group planting with shrubs beneath
Vase shaped as in *Delonix regia*	• high, wide-spreading branches • Majestic appearance • Gives excellent shade • An uncommon tree shape	• avenue trees • frame structures • used above large shrubs or small trees
Pyramidal as in *Araucaria excelsa*	• geometric in shape • pyramidal shape less noticed as the tree matures	• accent plant • avoid planting large trees near small buildings
Round as in *Mangifera indica*	• width and height uniformly distributed at maturity • usually dense foliage	• avenue tree, lawn tree • suitable for line planting, group planting • for enframement
Columnar as in *Millingtonia hortensis*	• rigid in appearance • vertically distributed	• accent plant • frame views and structures • useful in formal settings • line planting, group planting

5
Avenue Planting

What is the use of several trees grown in the forest? A single tree by the way side is far better where under people can rest (Vrikshayurveda).

The towns and villages are made more healthy, more beautiful and more efficient by well planned streets with avenue planting. There is no doubt that, an impressive avenue gives any property the special distinction that it desires. In the past, Emperor Ashoka and Mughal rulers paid attention to roadside planting. Avenue trees primarily provide shade and act as windbreaks, weather conditioners-by purifying and humidifying the air, check radiation of heat and obstruct glare, limit noise and enhance the beauty of the street where they are planted. Roadside planting has two aspects; beauty and utility and both these aspects should be combined. This requires a thorough knowledge of plant material and imagination in planning.

Roads are the important sites of the urban areas which contribute significantly in mitigating pollution by planting trees on both sides. For this we have to study the types of road, overhead electrical cable, space available on both sides, central verge, traffic triangles, roundabout, squares and other open spaces available before taking up any planting. It has been observed that trees, which are drought or frost resistant are generally tolerant to pollution. Ornamental dense foliage trees to provide good shade and flowering trees with pleasing colours are often used for avenue planting. Trees which adapt to the prevailing conditions, quick growing and which live for more number of years are preferred. Emphasis should be given to the native plant species which are comparatively well acclimatized, stress and pollution tolerant. The following factors should also be taken into consideration.

- Agro-climatic suitability
- Height and spread
- Canopy architecture
- Growth rate and habit
- Aesthetic effect
- Pollution tolerance and dust scavenging capacity

Designs of planting

Roadside land is more available in rural areas whereas in the city the land is limited and of private or government ownership. In urban areas, avenue planting may also be complicated by services such as gas, electricity, water, pedestrian and vehicle routes. There are many designs which can be adopted for planting trees in urban avenues (Randhawa, 1978). The design can be broadly classified into random planting, formal planting and informal planting.

Random planting

Trees are planted randomly without proper selection and without a proper design. This lacks proper planning. Many types of trees may be found on the road berms in any position. This type of planting lacks harmony and gives rise to a feeling of disorderliness.

Formal planting (Line planting)

Trees are planted in a definite pattern almost always in lines parallel to roads. The pattern is sub-divided into homogenous planting and heterogeneous planting. In homogenous planting a single variety is planted in lines parallel to the edge of the road. The design gives a definite character to the road and the roads can be named after the tree species used for planting. The planting may be in a single row with trees equidistant from each other on one side of the road only. This is suitable for narrow roads and when overhead wiring and sewers are located on one berm of the road. To provide shade for greater part of the day, roads running north south should have the trees on the western side and when they run east-west, the trees should be on the southern side.

In wider roads and where space is available, the trees are planted on both sides. This is suitable where symmetry and uniformity of size and colour are desired. The trees may be directly opposite or diagonally opposite to each other.

In heterogenous planting, trees of different species and varieties are used. Pleasing contrasts of colour, form and foliage are obtainable by planting trees of more than one variety. Trees with contrast in foliage texture, flower colour and form can be used together. Some of the good combinations are: *Terminalia* and *Tamarindus indica*, *Albizia procera* and *Mimusops elengi*, *Casuarina equisetifolia* and *Ficus lacor*. Flowering trees which flower at the same time and with blending colours such as *Peltophorum ferrugineum* and *Delonix regia*, *Cassia fistula* and *Delonix regia*, *Peltophorum* and *Colvillea*, *Jacaranda* and *Grevillea* are good combinations. In this arrangement single-row, double-row or multiple-row planting are also possible. Spreading and tall-growing trees are alternated, the latter to provide skyline or to harmonize with tall buildings and the former to provide shade. In multiple-row planting, inner rows may be tall ones and outer rows with spreading trees or inner flowering and outer evergreen. Belts of trees with different colour combinations can also be planted.

Informal planting

In this type of planting, more than one kind of trees and shrubs are used to decorate country roads or roads on hillsides. Roads may be rural and urban ones. Those passing through the countryside belong to the former group whether they are national or state highways or others. Due to their diversity in characters they need different treatments. Planting may be done to have brilliant flowering trees for beauty and shade or just to merge with the flora of the countryside. Utility trees to provide timber, fruit or fuel can also be used.

Where rural roads are broad two rows of trees may be planted 9 m from the centre of the road and 12-15 m between trees. *Albizia lebbek, Bombax ceiba, Azadirachta indica, Peltophorum ferrugineum, Syzygium cumini, Tamarindus indica, Kigelia pinnata,* etc., are suitable for this purpose.

Intermediate one with 8 m of road berms may have a single line of trees like *Albizia procera, Alstonia scholaris, Madhuka indica, Cassia fistula, Erythrina indica, Millingtonia hortensis,* etc.

On narrow roads only slender trees such as *Azadirachta indica, Milletia ovalifolia, Morus alba, Hardwickia binata* and *Pongamia pinnata* may be planted on the wider berm.

A well designed planting system is required before laying out of the roads. Urban roads pass through different sectors such as residential, shopping and industrial areas. Imaginative planning is required depending upon width of the road and the space available for planting the area. Where the street passes through open areas, large trees such as *Tamarindus, Terminalia, Peltophorum, Cassia siamea* and *Kigelia* can be planted. On the other hand, in residential areas, trees with smaller spread are required. *Terminalia arjuna, Alstonia scholaris, Polyalthia longifolia, Grevillea robusta, Lagerstroemia speciosa,* etc. are suitable. Fruit trees can also be used in such areas.

The importance of planting trees in different combinations had been emphasized in the *Vrikshayurveda*

Karamarda (Carissa carandas) and Vamsa (Bambusa arundinacea) to the east, paravata (Cicca acida) trees to the south, badari (Ziziphus mauritiana), kapittha (Limonia acidissima) and dhatri (Phyllanthus emblica) to the north are auspicious. Other trees having fibrous (or flexible) roots and of good, mediocre, and bad categories also should be planted on the border along with the trees of the same category at regular intervals.

The fruit-yielding and flower-yielding trees should be at the centre and other trees around them. They should be planted in pairs and they too, should be encircled by a trench.

Colour scheme for avenues

Most of the flowering trees look beautiful when planted in pure avenues. There are some species which flower at the same time, the colour of their flowers also harmonize, and hence they appear more effective when planted side by side. Some of the flowering trees which flower in the same season are grouped below in schemes with due regard to colour harmony and are recommended for planting along our town roads. Randhawa (1978) suggested striking colour schemes for avenues with flowering trees.

The colour scheme with *Cassia fistula* (golden yellow) - *Delonix regia* (scarlet-orange) - *Cassia fistula* (golden yellow) is very striking. The rich yellow colour of *Cassia fistula* flowers contrast with the scarlet-orange colour of *Delonix regia* flowers in the month of May when both the trees flower.

The colour scheme with *Peltophorum ferrugineum* (golden yellow) - *Colvillea racemosa* (orange red) - *Peltophorum ferugineum* (golden yellow) is very effective in October when both these trees are flowering.

The colour scheme with *Jacaranda mimosaefolia* (blue) - *Grevillea robusta* (yellow) - *Jacaranda mimosaefolia* (blue) is very attractive. These trees flower together in April and a beautiful colour effect, which is soothing in the glare of the April sunshine, is produced.

Another colour parade of flowers can be obtained by planting *Spathodea companulata* (orange crimson) - *Erythrina indica* (scarlet red) - *Spathodea companulata* (orange crimson). Both these trees flower in March and they are in a blaze of colour.

An attractive combination of *Cassia nodosa* (pink) - *Cassia marginata* (brick red) - *Cassia nodosa* (pink) which flower in May and June produce a very pleasing colour scheme of pink and brick red.

The colour scheme which is composed of three varieties of *Bauhinia variegata*, viz., *Bauhinia variegata* (purple mauve) - *Bauhinia variegata* (white) - *Bauhinia variegata* (light pink) is recommended for dust-free roads of residential areas. All these trees blossom in a leafless condition from middle of February to middle of March when they look like huge bouquets of pink, white, purple and light magenta flowers.

Naming roads after trees

Bio-aesthetic planning will also simplify the problem of naming roads and streets. Streets can be named after the flowering trees, which are grown on the road, such as golden shower avenue (*Cassia fistula*), gul-mohar avenue (*Delonix regia*) and asoka avenue (*Saraca indica*). Not only will the roads be readily recognizable, but this device will also enable the citizens to familiarize themselves with our common flowering trees. Some imaginative pioneer has actually named a road in Lucknow as 'Millingtonia' Avenue after *Millingtonia hortensis.* All the roads in Punjab Agricultural University, Ludhiana, are named after flowering trees which are grown in them in the form of avenues.

In every big city, we find triangular pieces of land at the junction of roads. To safeguard against traffic accidents, these plots are not leased for building purposes. There are many such triangular plots in these cities. At present, they are lying neglected, covered with ugly, wild-growing jungle trees. How beautiful they can be, if planted with flowering trees. Trees of only one species should be planted in each triangle. Covered with spathodeas, lagerstroemias, jacarandas and *kachnars*, these triangular plots will appear very beautiful, serving as nodal points of beauty, affording a welcome shade to the pedestrians. Beautiful roads and well-planned parks and squares will bring the beauty of nature within the reach of the common man in our towns and cities.

Avenue of Honour

Practice of planting of trees for commemorative purposes dates back to antiquity, the Romans pioneering the concept of commemorative tree planting along roads with burying their dead outside the city in tombs strung along the roadside.

In Australia, commemorative trees have been planted in public spaces since the late nineteenth century. Numerous trees were planted in parks in Melbourne and throughout Victoria to mark the visits of important and famous people. This tradition of commemorative planting was continued in 1901 when at the end of the Boer War trees were often planted for each soldier of the district who was killed in South Africa. In Australia, the use of trees as a memorial to soldiers may date from the Boer War, however these plantings rarely consisted of more than two or three trees in each town. Around 121 such Avenues of Honour were planted throughout Australia in response to the First World War. The Avenue of Honour, located along the Ballarat Burrumbeet Road, Ballarat is presently comprised of a total of 3,332 trees, and covers a distance of approximately 22km in length. The first 1000 trees of the avenue were planted in 1917. The trees were planted in single lines along either side of the road at a regular spacing of 10-15 m apart, and set back from the edge of the carriage way approximately 5-6 m. Each newly planted tree was protected by a substantial timber guard, to which a plate bearing each soldier's name, rank and unit was attached. Originally 23 different species of trees were planted in the Avenue. Individual species were usually planted in blocks of about 50 trees (25 either side), however during the last two sections of the Avenue, a slight change was made and two different species were used alternately and planted in blocks of around 100 trees. Many of the original species used in the Avenue did not flourish and were soon replaced by several different species of Elm, *Ulmus sp.* and poplar, *Populus sp.*

Planting along railway lines

The vacant land along the railway lines provides a very good opportunity for tree planting. It will help to transform the land into a colourful place and travellers will carry back happy impressions and a long tiresome journey will be greatly relieved of its boredom.

Usually trees of economic importance may be preferred for planting along railway lines. The ornamental aspect should also be considered. Watering the trees on railway tracts is not always possible and the seedlings may have to depend on rains after a certain stage. As such, trees, which can withstand drought, should be preferred in dry localities. The selection would, however, depend upon the climate. They should, however, be long-lived and deep-rooted. Shallow rooted ones may be blown over by wind during storms and many may obstruct the traffic. Trees with spreading branches are not suitable since they require periodical pruning. Shade trees along railway lines will also offer a resting-place for gang workers. The planting should be done leaving the specified distance as required by railway rules.

Palms and Cycads for avenue planting

Palms and cycads are especially useful in the development of the tropical effect in landscape beautification. The true palms all belong to the botanical family Arecaceas. The cycads, including the Sago palms and the Zamia, belong to family Cycadaceae. Palms of the hardier types will survive in many parts of the country.

The stately fan palms and the graceful *Phoenix* and *Cocos* palms may be used for avenue planting, in groups at the end of roads, or to give height to mass plants of the more dwarf types of plants. The dwarf forms such as *Cocos australis, Phoenix roebelenii* and *Cycas revoluta* are especially useful for small group plantings.

All palms bloom although the cream coloured bloom is relatively inconspicuous and therefore unimportant. Pruning is an important job for care of this group of plants.

Description of palms

Chrysalidocarpus (Areca) lutescens. Golden Feather Palm. A palm useful for porch or patio planting. It produces a slender, ringed trunk and graceful curving pinnate foliage with golden yellow stems. Suckering freely at the base, it produces a bushy effect. This palm is very tender to cold.

Cocos australis. Dwarf Blue Palm, Australian Palm, Pindo Palm. A rather dwarf palm which produces graceful, recurved, blue-green, pinnate leaves. The bright yellow, edible fruits are borne in heavy, ornamental clusters. This palm is well adapted to sub tropical conditions.

Cocos nucifera. Coconut Palm. Although this palm is very valuable for its ornamental pinnate foliage and its edible "nuts," it is too tender to cold to be grown in this region except in sheltered locations along the coast.

Cocos plumosa. Queen Palm, Plumy Coconut Palm. A tall, slender palm which produces a smooth light gray trunk and graceful, feathery, dark green leaves. It is tender to cold and occasionally chlorotic. This palm can be grown successfully in the Lower Rio Grande Valley in sheltered locations.

Cycas circinalis. Queen Cycad, Fern Sago. A large cycad which attains a height of 2 to 3m. It is symmetrical in its habit of growth, producing graceful, bright green, pinnate foliage. The foliage is subject to sun scald. It is recommended that this plant be given a shady location.

Cycas revoluta. Dwarf Sago Palm. A dwarf plant of stiff, symmetrical appearance. It produces very dark green, stiff, pinnate leaves which have commercial value in floral decorations. Well adapted, and easily propagated by offshoots.

Erythea armata. Blue Fan Palm. A very slow growing palm whose stiff bluish-gray "fans" are covered with a powdery bloom on both surfaces.

Howea belmoreana. Kentia Palm, Curly Palm. A small palm useful for porch or patio. It produces slender, graceful, dark green pinnate leaves at the top of a small, upright trunk.

Latania commersonii (Livingstonia chinensis). Patio Palm, Chinese Fan Palm. A dwarf fan palm which produces bright green, fan shaped leaves and a slender dark brown trunk. This palm is suitable for porch or patio use only, since it is very tender to cold and subject to wind burn.

Oreodoxa (Roystonea) regia. Royal Palm. A tall, slender palm which produces a smooth light gray trunk and a head of graceful, bright green, pinnate leaves. This palm is too tender to cold to be used except in a sheltered location.

Phoenix canariensis. Ornamental Date Palm. A large, spreading palm which produces drooping, dark green, pinnate leaves and ornamental fruit clusters. It is well adapted and useful in specimen, group or avenue plantings.

Phoenix dactylifera. Date Palm. A large spreading palm which produces gray-green, pinnate leaves that are rather stiff and upright, giving the plant a less pleasing appearance than the ornamental date. However, it is well adapted and can be used where a large specimen plant is needed. If edible fruits are desired, it is recommended that offshoots from superior fruiting plants be secured.

Phoenix reclinata. Leaning Palm. A spreading *Phoenix* palm somewhat resembling the ornamental date but having a tendency to form bushy clusters of offshoots. The palm presents a better appearance if the offshoots are allowed to remain. This plant has proven to be quite slow growing and somewhat tender to cold.

Phoenix roebelenii. Pigmy Date Palm. A very dwarf palm which produces a small, upright trunk, and a head of very fine, fern-like foliage. It is quite tender to cold, and is the most useful in patio plantings.

Sabal blackburnia. Blackburn Palmetto Palm, Giant Palmetto Palm. This is the larger growing species of palmetto which is native to other parts of the state.

Sabal texana. Texas Palmetto Palm. The Sabal palms are well adapted and should be extensively used. The native *S. texana* is an upright fan palm producing blue-green foliage. The trunks of the young palms are the most attractively laced or "booted" with the green leaf petioles which remain firm and green on the palm trunk for several years. This palm is highly resistant to wind and cold damage. It is the first palm tree sighted by Europeans on the North American continent, and was responsible for the first naming of the Rio Grande, "Rio de las Palmas."

Trachycarpus fortunei (Chamaerops excelsus). Windmill Palm. These upright fan palms seldom reach a height of over 3 m. The "fans" are set windmill-fashion on the trunk. The trunk is covered with coarse, loose, brown fiber.

Washingtonia robusta and *filifera.* Washingtonia Fan Palm. Both of these species of tall fan palms grow in this region, and are well adapted. *W.*

robusta, the taller species, is the most commonly planted. The leaves are cleft 2/3 to the base, and the petiole margins are completely armed. *W. filifera* produces leaves which are cleft almost to the middle and the petiole margins are armed about half the length. This species is more wind resistant.

Zamia integrifolia (Zamia), *Z. floridana* (Coontie) and *Z. pumila.* (Comptie). These dark green, dwarf cycads are very ornamental, being especially useful as foreground subjects in palm plantings. There are several species of Zamias native to Florida, some of which are collected by the Indians as a source of food and income. The fleshy underground roots produce the arrowroot of commerce.

Table 4. Palms and Cycads for avenues

Name	Mature size		Notes
	Height (feet)	Spread (feet)	
Chrysalidocarpus lutescens Golden Feather Palm	10-15	3- 4	Useful as indoor subject
Cocos australis Pindo Palm	8-10	6- 8	Excellent dwarf palm
Cocos nucifera Coconut Palm	20-25	15-20	Useful only on the coast
Cocos plumosa Queen Palm	20-25	10-15	Useful as specimen
Cycas circinalis Fern Sago	6-15	6-10	A shade loving, dwarf, palm-like plant
Cycas revoluta Dwarf Sago	2- 6	5- 6	A symmetrical, dark green, dwarf plant
Erythea armata Blue Fan Palm	10-20	4- 8	Desirable, small, fan palm
Howea belmoreana Kentia Palm	3	2	For indoor use
Latania commersonii Patio Palm	5-15	5-10	Indoor use only

Contd...

Ornamental trees suitable for town roads

Foliage trees for outer avenue	Flowering trees for inner avenue
Averrhoa carambola	*Cassia fistula*
Callistemon lanceolatus	*Bauhinea-purpurea, B. variegata*
Anthocephalus cadamba	*Colvillea racemosa*
Eugenia operculata	*Peltophorum ferrugineum*
Polyalthia longifolia	*Spathodea nilotica*
Putranjiva roxburghii	*Jacaranda mimosaefolia*
Sterculia alata	*Poinciana regia* (also *Delonix regia*)
Pithecolobium saman	*Lagerstroemia flos-reginae*
Melia azadirachta	*Grevillea robusta*
Tamarindus indica	*Gliricidia maculata*

Trees for narrow roads

- *Calophyllum inophyllum*
- *Cassia javanica*
- *Cassia nodosa*
- *Cassia renigera*
- *Casuarina equisetifolia*
- *Diospyros peregrina*
- *Filicium decipiens*
- *Grevillea robusta*
- *Jacaranda mimosaefolia*
- *Kigelia pinnata*
- *Lagerstroemia flop-reginae*
- *Melia azadirachta*
- *Millettia ovalifolia*
- *Emblica officinalis*
- *Putranjiva roxburghii*

Trees for wider roads

- *Colvillea racemosa*
- *Ficus lacor*
- *Ficus retusa*
- *Mimusops elengi*
- *Polyalthia longifolia*
- *Pongamia pinnata*
- *Samanea saman*

Trees for providing colour and aesthetic look on the road

i) Ornamental flowering trees used for small roads

- *Amherstia nobilis*
- *Bauhinia acuminata*
- *C. renigera*
- *Gliricidia maculata*

- *B. variegata*
- *Brownea grandiceps*
- *Cassia fistula*
- *C. javanica*
- *C. nodosa*
- *Jacaranda mimosaefolia*
- *lagerstroemia speciosa*
- *L. thorellii*
- *Saraca indica*
- *Spathodea campanulata*

ii) Ornamental trees for wider roads

- *Bauhinia triandra*
- *Bombax triandra*
- *Bombax ceiba*
- *Colvillea racemosa*
- *Erythrina indica*
- *Peltophorum inerme*

6

Treescapes An Ecofriendly Approach

Urbanization coupled with industrial development creates problems of environment like air pollution, water pollution, noise pollution and civic ugliness. Of late, there is a greater awareness of environmental issues and increasing understanding of several attributes of vegetation. This led to an intensification of activity in the field of landscape horticulture to combat pollution and create an aesthetic environment for recreation and healthy living. Planning and execution of landscape projects for different urban locations should be done to meet both aesthetic and ecological requirements.

Psychological effect of plants

For centuries, it was assumed, both in Western and Eastern urban civilizations, that visual contact with plants or other natural elements might stimulate psychological well being and has beneficial effect on man. The first documented political interest in people's need for greenery dates back to the time when ancient Rome became urbanized. As the city became noisy and overcrowded, the inhabitants complained that life became very stressful. They found that the countryside or the gardens of Rome were the only place where they experienced complete recreation.

In terms of psychological aspects of our health and well-being, modern lifestyle also exposes man to new situations. Studies of how the environment and our surroundings affect us mentally are called environmental psychology. Studies in this field showed clearly that well-being and levels of psychological and physiological stress are significantly influenced by our surroundings. Today, plants are widely used in urban environments, both outdoor and indoor. Scientific studies could bring forward convincing evidence regarding beneficial effects of plants on human health and well-being. As a result use of plants both outdoor and indoor could receive an additional dimension to their value as pure ornamentals.

Pollution control

Environment is degrading rapidly due to various kinds of pollution. Environmental pollution is an important global problem. Fresh air, pure water and noise free surroundings are the main criteria for an ideal environment.

Pollution monitoring reveals that 35-40% of the pollutants of the atmosphere is composed of particulate matter in India. Plants can filter out dust, smoke and noise. Among several measures, that have been taken up against air pollution, landscaping with pollutant tolerant trees in and around industrial areas and metropolitan cities is one of the most effective measures, providing aesthetic value as a bonus. Shape of the leaf, presence of leaf hair, phyllotaxy, age of leaf and type of plant associations are the determining factors, for dust and noise collecting efficiency.

Atmospheric gases

Gaseous pollutants found in the atmospheric air are Sulphur dioxide, Ozone, Peroxy acetyl nitrates (PAN), Hydrogen fluoride and Oxides of nitrogen. Vehicles are major sources of nitrous oxide and hydrocarbons, which cause high level of ozone. Generally enhancement of respiration occurs with high doses of air pollutants and this results in wasteful use of carbohydrates and energy, which is normally used for plant growth. Uptake of gaseous pollutants affects morphological features such as stomatal frequency, density and cuticular permeability. Tolerance and susceptibility of plants to a particular pollutant should be studied thoroughly before going for planting against environment pollution. Under high level of ozone, stomata remain closed and lead to accumulation of harmful chemicals. Plants with high stomatal conductivity were susceptible to gaseous air pollution.

Sulphur dioxide

Sources of sulphur dioxide pollution are refining of Sulphur containing Cu, Pb, Zn and Ni ores, use of high sulphur petroleum and natural gas and use of low quality and high sulphur containing coal.

As all gaseous exchanges are taking place through stomata, which are micropores, because of air pollutants the regulation of opening and closing system of stomata is disturbed. So_2 once enters into the cell, reacts with proteins of plasma membranes. Sulphur dioxide is converted into sulphite and bisulphite ions. These sulphite ions are highly toxic than sulphur dioxide. Chemically and biochemically transformation of sulphite plays a very active role in making the plant susceptible or tolerant, as it is a detoxification process requiring high energy.

The most sensitive plants are affected by suplhur dioxide even at a lower concentration of 0.05 ppm. Moderately sensitive plants withstand 0.3 ppm level. But the immune plants are quite unaffected by exposure at 3 ppm level for a few hours.

Plants with high amount of ascorbic acid content should be planted against SO_2 pollution. Ascorbic acid detoxifies SO_2 and the plant system is not much disturbed. Presence of enzymes such as sulfite oxidase helps in SO_2 tolerance. Trees resistant to SO_2 pollution are *Albizia lebbek, Ailanthus excelsa, Alstonia scholaris, Azadirachta indica, Ficus religiosa, Lagerstoremia flos-reginae, Polyalthia longifolia, Terminalia arjuna, Quercus palustris* and *Q. rubra.*

Ozone

Under high level of ozone, stomata remain closed which lead to accumulation of harmful chemicals. Like SO_2 tolerant plants, plants having high ascorbic acid content are tolerant to ozone pollution. Presence of enzymes such as cationic peroxidases is responsible for ozone tolerance. Plants resistant to ozone pollution are *Acer campestra, Acer plantanoides, Acer negundo, Euonymus europaeus* and *Quercus rubra*.

Peroxy acetyl nitrate (PAN)

Sources of PAN are secondary products of photo-chemical reactions between nitrogen dioxide and simple olfeins. PAN resistant trees are *Acer plantanoides, Acer negundo* and *Quercus palustris*.

Hydrogen fluoride

Sources of this pollutant are fertilizers, cryolite bath, phosphate fertilizer plant, refining of aluminium ore, brick plant, steel manufacturing plant etc. Hydrogen fluoride resistant trees are *Acer campestra, Euonymus europaeus, Juniperus sp.* and *Quercus sp.*

Oxides of Nitrogen

Oxides of nitrogen appear in atmosphere from fertilizers, fossils, fuel combustions and biomass burning. Resistant trees are *Pinus sp., Robinia pseudocacia, Quercus rubra, Sambucus nigra, Taxus baccata* and *Ulnus montana*.

Dust and smoke

Trees having compact branching and closely arranged broad, elliptical leaves with cuticular hairs collect a lot of dust from atmospheric air. Leaves with shiny surface or waxy in nature will collect dust which is easily washed away by rain water. Dust collection ability of trees depends on certain morphological characteristics like branching habit, arrangement of leaves, size, shape, surfaces (smooth/hairy) and presence or absence of trichomes. Compact branching and arrangement of leaf increase dust-collecting ability of trees. Planting green belt of trees of various combinations controls dust pollution. Width of green belt and density of plantation depend on source and intensity of dust pollution. Dust fall can be reduced 2 to 3 times by 8 m wide green belt between road and buildings. A reduction of overall dust fall upto 42% was noticed by planting conifers in temperate urbanized dwellings. Suitable tree species for controlling dust pollution are *Acacia auriculiformis, Alstonia scholaris, Bombax ceiba, Butea monosperma, Cassia siamea, Dalbergia sissoo, Ficus benjamina, Ficus benghalensis, Ficus infectoria, Madhuka indica, Mangifera indica, Peltophorum ferrugineum, Polyalthia longifolia, Pongamia glabra, Shorea robusta, Syzygium cumini, Tectona grandis* and *Terminalia arjuna.*

Noise

Trees play a very important role in reducing noise pollution by absorption, deflection, reflection, refraction and masking. The branches, twigs and leaves of trees absorb sound waves. Some parts of trees are flexible and light having many thick and fleshy leaves. The sound is also deflected and refracted by heavier branches and trunk of trees. Main phenomenon involved in reducing noise pollution is by introduction of trees between source and receiver, which reduces sound wave frequency and intensity.

According to speed of vehicles, various schemes of planting trees have been proposed. Against high speed vehicles, planting small and large tree green belts of 18 to 30 m width, 15 to 27 m from traffic lanes is recommended. In such scheme, trees in central rows should be at least 13 m tall. For moderate speed vehicle in urban areas planting of small and large tree green belts of 6 to 15 m width which are very effective, 6-15 m from the centre of the nearest traffic lane. Shrubs of 1.8 - 4.0 m height should be used next to the traffic lane followed by backup rows of trees of 4.5 to 9.0 m tall.

While planting against noise pollution, some of the important considerations are selection of plant species and their arrangement in relation to sound source and receiver, planting height and planting density, wind direction and wind velocity, temperature and humidity of particular area. In addition to deflection, absorption, reflection, refraction and masking, plants absorb sound by their influence on local climate by modification of wind speed and stabilization of temperature and humidity. Where year-round noise screening is desired, evergreen species are to be planted. To form a dense and continuous barrier, trees and shrubs should be planted as close together as possible.

Plants having thick and fleshy leaves with flexible petioles and capacity to withstand vibration are suitable for planting against noise pollution. Heavier branches of tree and trunk of tree also deflect or refract sound waves. *Alstonia scholaris, Acer negundo, Azadirachta indica, Butea monosperma, Betula pendula, Erythrina indica, Eucalyptus citriodora, Gravillea robusta, Juniperus chinensis, Kigelia pinnata, Mangifera indica, Madhuka indica and Terminalia arjuna* are suitable against noise pollution.

Lead

Lead deposited on plants enters through stomata. Plant proteins are rich in amino acids such as cystein. Lead, cadmium and zinc bind with these amino acids and environment becomes free from these pollutants.

Herbs with large leaf area such as lettuce or with extensively dissected foliage (parsley) accumulate high amount of lead when grown near urban roads and near factories. Trees with hairy foliage are not suited for lead pollution since such deposition of lead contaminates wood and surroundings. Protective planting and green belts diminish lead content in air. *Cassia siamea* and *Zizypus mauritiana* are resistant to lead pollution.

Trees work against the pollutant by active filtration, protection of sedimented pollutants against pick-up by wind and by passive diversion of airflow. However, filtration depends on meteorological condition, topography, kind and concentration of pollutants, position of green belts and the structure of green belts. Dust filtering effect depends on physical, chemical and physiological properties of plant species. Filtering effects are better in coniferous trees than that of deciduous trees. In compact hedges, sedimentation of dust is more in central and basal regions, while in loose hedges large sedimentation is on the apical region of plant. Dispersion of atmospheric pollutants depends to a large extent on nature of planting, which provides screening. Thin screens created by planting rows of trees generally divert and redirect moving air. Green belts, heavily wooded areas and cluster of trees create thick screens, which tend to filter masses of air and accumulate particulate or gaseous pollutants on their foliage.

Pollution indicator tree

Couroupita guianensis (Cannon ball tree) regularly responds to seasonal climatic changes. The phenomenon was observed repeatedly. At the onset of monsoon, this is the first tree to start shedding leaves. It also sheds its leaves with onset of winter. But this tree sheds all its leaves in two hours due to bursting of a chlorine cylinder and starts sprouting afresh within 24 hours. These indicators may also prove a boon in legal cases usually filed by inhabitants or farmers surrounding an industrial estate.

Bioallergen trees

Bioallergens are responsible for health problems in and around cities. Pollen from various trees causes disorders such as asthma and dermatitis. It is stated that *Holoptelea integrifolia* and *Prosopis juliflora* are two important sources of aeroallergens for patients residing in Northern Western and in peninsular region of India. Pollen grains of *Anacardium occidentale, Borassus flabellifer, Casuarina equisetifolia, Cocos nucifera* and certain other palms are important sources of aeroallergens in coastal and peninsular regions.

Pollen from *Ailanthus* excelsa, *Albizia lebbeck, Broussonetia papyrifera, Cassia spp., Etretia laevis, Eucalyptus* spp., *Holoptelea integrifolia, Kigelia pinnata, Morus alba, Prosopis juliflora, Putranjiva roxburghii, Salvadora persica* and *Melia azadirachta* are reported to cause allergic reactions in specific cases.

Legumes in landscaping

In 21st century, more thrust is given to legumes due to their biological nitrogen fixation and multifarious uses as fuel, fodder, luxury timber and good ornamental. Fast growing nature encourages landscapers and forest officials to choose legumes in urban tree planting programmes

Common multipurpose leguminous tree species, which serve the above purposes, are *Acacia species, Dalbergia sissoo, Leucaena leucocephala, Samanea saman* and *Xylia xylocarpa.*

Work has to be done to screen more multipurpose legume species which should suit to urban plantation programme. There is also need to intermingle agro-forestry programme with urban tree planting programmes.

Trees for various urban locations

Industrial areas and factories

More recent studies from Germany and Australia confirmed ability of plants and the ecosystem of plants and microorganisms to be possible powerful air purifiers, even if concentration of chemicals is low. This air-purifying ecosystem needs a period of time approximately 14 days to adapt to chemicals that occur in air. System seems to be independent to light - which means that photosynthetic activity of plants is less important than symbiotic activity of roots and microorganisms at least after the system has become adapted to environment.

The trees planted in industrial areas should be tolerant towards pollutant and should have dense, broad, foliage with slight rough texture for trapping dust. Plants selected remain effective in abatement of pollution and revitalization of environment for a long period of time.

Parks and other recreation places

These are places of relaxation, entertainment and getting fresh air in an otherwise congested city. Good flowering and shade trees should be planted in groups or singly in some corner (as focal point) or other suitable places for creating beauty as well as a place of rest. Accent trees with attractive foliage or flower can also be used in parks. Development of 'Theme parks' creates awareness among citizens regarding different plant species.

Public buildings and educational institutions

Recent studies concluded that if office personnel could view greenery through their office windows, significantly less stress was reported during their working days compared with office workers who had a view to non-vegetated areas, like streets and parking lots.

General recommendation is to plant large majestic looking trees, which provide considerable shade and also to keep down severe heat and cold. Trees planted should not completely obstruct view from outside. For enhancing beauty, it is suggested to plant a row of flowering trees, with different blooming seasons, in front of large trees along periphery.

Hospitals

Several studies show that even the view through windows may affect our well being and health. In a hospital study, beneficial influences on patients of a natural view from a bedside window were considered. Recovery data after surgery was compared from two groups of patients that were essentially identical apart from their window view: one group of patients looked out onto trees, whereas the other had a view of a brick wall. Patients with a 'green' view had shorter post-operative hospital stays, required fewer injections of strong pain reducing drugs and tended to have fewer minor post-surgical complications such as persistent headache or dizziness.

Flowering trees are always preferable in hospitals. Trees producing pleasing light coloured flowers of rose, white, pink and other partial shades could produce a feeling of harmony to patients. Dark and vibrant coloured flower producing trees especially red coloured tree like gulmohar (*Delonix regia*) may be avoided. Shade trees may also be planted in hospitals, in places like parking areas.

Star hotels

Environment is increasingly being seen as handmaiden of tourism. Eco-friendly hotels that are rooted in environment are the need of day. In star hotels, planting of trees is mainly done to have a link with surrounding environment, for screening for privacy and for aesthetic value. Line planting of trees with wide oval canopy to screen living areas and swimming pools is recommended. To add to aesthetic value ornamental trees with bright foliage and flowers can be used.

Religious places

Religious places like temples, churches and mosques are places suitable for planting trees. Trees like *Ficus religiosa, Saraca indica* and

Plumeria alba are associated with temples from ancient period. Flowers of specific trees are used for offerings in temples. In churches and mosques, flowering and foliage trees which merge with serene atmosphere are used.

Home garden

In houses having a restricted space, the plan may be to build around a square or rectangular lawn to give a spacious effect. Complicated geometric designs need not always be adopted and a sort of naturalistic planting with gentle curving lines may be easy. Space available should be put to maximum use depending upon other needs such as a play ground for children, garage, drive etc. Shaded area, if any, can be used as utility area and open sunny area may be used for planting. Kind and number of plants should be limited. A good fence may be needed, if there is no compound wall. A hedge or a tree with dense crown serves the purpose of screening windows from neighbouring houses. Utility crops such as vegetables and one or two fruit plants have to be accommodated in plan. If there is a space in backyard, it may be used for vegetables and a kitchen garden can be raised. Rest of the area should be left for ornamental plants. In small houses, there may not be enough space for large trees. One or two medium sized trees and more of shrubs, climbers and annuals may fit in well. Ornamental trees should be planted only on outer boundaries. It is no use planting avenues on inner roads in a small compound. Only dwarf flowering trees should be grown avoiding use of large trees with spreading umbrella-like crowns such as *Delonix regia*. Each locality requires to be planted with suitable species. If space is available, trees with fragrant flowers can be planted nearer the house. Trees, which yield edible fruits, are also useful to grow as they attract a variety of birds also and make the garden a more interesting place.

City forests and open green belts

At present our densely populated urban areas are treeless and bare. Use of trees in human settlements in form of groves in rows, or other settings enhances beauty of landscape. Plants are widely used in urban environment wherever space is available. Group planting of trees and shrubs and creation of open green belts with single or different kinds of trees is to be included in modern city planting. Group planting of different kinds of trees and other plants in cities creates impression of a small forest.

Planting trees in villages

In villages, we require trees, which provide fuel and fruit as well as timber. Species selected must be fast growing and easily grown. Fruit trees like mango, tamarind and jamun may be selected for planting in villages. Fodder trees provide valuable cattle feed in winter/summer when grasses are not available. As compared with grasses, some leaf fodder, particularly those of *kachnar* and *toot* are exceptionally rich in crude fat and protein, and nitrogen free extracts. Good leaf fodder species are *kachnar, toot, neem* and *babul*. Bamboos can easily be planted near ponds. Bamboo has many uses in the villages.

Wild fruit trees like *ber* (*Zizyphus jujuba*) and *toot* (*Morus alba*) can be propagated on wastelands and in forests on a large scale. *Ber* provides a delicious fruit of many varieties, eaten both in fresh and dried condition. It is also an excellent fodder tree and its leaves are fed to goats and buffaloes. It is an extremely hardy tree which can withstand both drought and frost, and is ideal for barren districts with comparatively poor rainfall.

Bioaesthetic planning

Bioaesthetic planning may be defined as conscious planning of flora and fauna with object of beautifying our surroundings. Credit of giving meaning and content to it goes to Dr. M.S. Randhawa, who in his inspiring book 'Beautifying India' gave an outline of a plan for beautifying landscape of India (Randhava, 1983).

Bioaesthetic planning embraces both animal and plant sciences, botany and zoology, and may further be defined as planned ecology of living beings from artistic and aesthetic points of view. Main objectives of bioaesthetic planning for India is the encouragement of planting of selected ornamental flowering trees in town and villages and preservation of graceful animals and harmless birds. For a bioaesthetic planner, the natural canvas is too vast and choice of colour too varied for drawing inspiration for creation of pleasing and salubrious effects.

Chandigarh experience

Chandigarh is one of the most carefully planned cities of world. It is situated at foot hills of Shivalika of Himalaya. City took shape in 1955 and credit for landscape planning, selection of green species, etc. go to late Mr. M.S. Randhawa - a well known plant scientist, artist, environmentalist and an administrator.

Urban Chandigarh is divided into 46 rectangular sectors plus an industrial area on the Southern side. Each sector has an area of approximately 100 ha. It is surrounded by 35 rural areas. The city has no natural forest. Likewise, there being no agriculture in the city, the agroforestry component is lacking. Nevertheless, the avenue forestry and garden forestry components are very strong (Kohli, 1996).

Tree planting in Chandigarh was done very carefully. Here selection of trees was done with consideration of aesthetic appearance and tree architecture, tree characters like leaf fall, shape and size of leaves and crown, colour and time of flowering, climatic adaptability and availability of healthy material

Chandigarh was planted with 99 types of trees which include 68 species of avenue trees, 11 species of other forest trees and shrubs in other gardens and 20 types of trees in the premises of city residents. City roads are mostly identified with type of trees along their sides. In general, city has six types of roads in addition to bicycle tracts. Avenue of heavy traffic automobile highway was planted with trees of light foliage. In busy market area, where urban activity is more, trees of deciduous and evergreen with more ornamental value were planted. City is also decorated with 11 gardens. Thus we find in Chandigarh that the town and country are blended, the result is harmony and link between nature and man is established. Chandigarh landscape provides a proto-type for development of several cities in India including Delhi and Bangalore.

Urban green area management

A common characteristic of urban landscape is intense use of land, since land is expensive and space is limited. This creates a compactness and sense of confinement in cities that can be oppressive. There are two aspects of tree space above and below ground, and both are important. Vertical space needs are more obvious in that trees should not be interfered with utility lines. This can be avoided by planting low growing trees or by regular pruning of canopy. The horizontal or underground space need to be aesthetically, greened. Root spread is important since it should not interfere with foundation of buildings. The horizontal reach of roots can be twice as wide as the tree length and trees must be given for more root space than have often been done.

Pavements are the only place available for growing trees in cities. But unfortunately, Municipalities in the name of beautification, are covering every inch of bare soil on pavements with cement concrete slabs. By doing so, the top fertile soil is removed, roots are ruthlessly cut and exposed. Exposed space is then packed with cement concrete, gravel and completely sealed with tiles, up to tree trunks.

Tiling is an extreme form of soil compaction that suffocates trees by preventing aeration and drainage of soil. Besides suffocating, tree roots also get charred due to excessive heating of tiles on a typical summer afternoon. In Delhi, indiscriminate and excessive tiling of pavements resulted in massive tree death. Different tree species take varying length of time to withstand any ill effect of tiling. *Alstonia, Bombax* and *Delonix* are not to withstand any disturbance to their roots. Hence concrete pavements should be restricted as far as possible. Pavements should be effectively utilized by planting suitable trees and should be maintained green and grassy.

General management techniques for urban trees

Management of trees in urban areas is very important, as there is lack of space for dwelling, playing, movement of traffic, etc. If not cared properly urban trees attain a gigantic size with irregular shape and thus might cause a lot of inconvenience to public. Keeping such considerations, urban tree management plans should include the following:

- Providing tree guards for young saplings
- Staking of saplings
- Training of plants
- Proper pruning technique
- Proper weeding and manuring
- Irrigation
- Tree surgery for treating dead wounds
- Proper pest and disease management practices
- Replanting old trees by new saplings

Recommendations for ecofriendly plantations

- Establishment of gardens in and around cities which should serve the purpose of education, recreation and conservation.
- Development of arboreta viz., botanical, medicinal, flowering and non-wood forest which help in conserving species.
- Comprehensive and co-ordinated implementation of multidisciplinary projects for establishment of sizable biosphere by various agencies.
- Future thrust has to be given for use of native trees, multi purpose legumes, conifers and fast growing species to provide global tree cover.
- Utilization of courtyards of houses, roads, public places, industrial areas, railway lines, parks, temple site etc. for planting trees.
- Attempts have to be made to include energy producing trees and timber yielding trees for getting economic benefits.
- Involve citizens voluntarily to manage urban green belts and gaining attention of administrators and town planners.
- Need for a comprehensive tree census for each urban city, planting plans and its execution according to industrial and vehicular density and population density.

- Need for consultative processes among city dwellers, Government, non-governmental organizations, landscapers etc. to make cities more livable.
- A close link between researchers and practitioners of urban landscaping.

Trees for various urban locations

1. Industrial areas and factories

a) Trees for reduction of gaseous pollution

- *Albizia lebbek*
- *Alstonia scholaris*
- *Azadirachta indica*
- *Euonymus europaeus*
- *Ficus religiosa*
- *Juniperus sp.*
- *Lagerstroemia flos-reginae*
- *Pinus sp.*
- *Polyalthia longifolia*
- *Quercus palustris*
- *Quercus rubra*
- *Terminalia arjuna*

b) Trees for reduction of noise pollution

- *Alstonia scholaris*
- *Butea monosperma*
- *Grevillea robusta*
- *Erythrina indica*
- *Lagerstroemia flos-reginae*
- *Pterospermum acerifolium*
- *Terminalia arjuna*
- *Mangifera indica*

2. Parks and other recreation places

a) For providing colour and aesthetic appearance

i) Ornamental flowering trees

- *Amherstia nobilis*
- *Bauhinia purpurea*
- *Brownea grandiceps*
- *Cassia fistula*
- *C. javanica*
- *C. nodosa*
- *C. renigera*
- *Delonix regia*
- *Jacaranda mimosaefolia*
- *Lagerstroemia speciosa*
- *Saraca indica*
- *Spathodea campanulata*

ii) Ornamental foliage trees

- *Caesalpinea coriarii*
- *Ficus benjamina*
- *Ficus benghalensis*
- *Ficus retusa*
- *Mimusops elengi*
- *Polyalthia longifolia*
- *Pongamia glabra*
- *Samanea saman*

iii) As accent plants

- *Amherstia nobilis*
- *Araucaria excelsa*
- *Brownea grandiceps*
- *Cassia nodosa*
- *C. javanica*

- *C. fistula*
- *C. renigera*
- *Callistemon lanceolatus*
- *Delonix regia*
- *Ficus lyrata*
- *Ficus krishnae*
- *Lagerstroemia flos-reginae*

3. *Public buildings and educational institutions*

a) Larger gardens or bigger compounds

- *Cassia renigera*
- *C. siamea*
- *Kigelia pinnata*
- *Mimusops elengi*
- *Peltophorum ferrugineum*
- *Polyalthia longifolia*
- *Pterospermum acerifolium*
- *Spathodea campanulata*

b) Maidans, recreation grounds, etc.

- *Caesalpinea coriarii*
- *Ficus benjamina*
- *Ficus religiosa*
- *Kigelia pinnata*
- *Mimusops elengi*
- *Polyalthia longifolia*
- *Pongamia glabra*
- *Thespesia populnea*

c) Wind break

- *Melia azadirachta*
- *Millingtonia hortensis*

- *Polyalthia longifolia*
- *Pongamia glabra*

4. Hospitals

a) Ornamental flowering trees

- *Bauhinia purpurea*
- *Cananga odorata*
- *Cassia nodosa*
- *Citharexylum spinosum*
- *Jacaranda mimosaefolia*
- *Michelia champaca*
- *Millingtonia hortensis*
- *Tabebuia rosea*

b) Ornamental foliage trees

- *Araucaria excelsa*
- *Ficus benjamina*
- *Ficus elastica*
- *Ficus lyrata*
- *Ficus religiosa*
- *Melia azadirachta*
- *Mimusops elengi*
- *Polyalthia longifolia*

5. Star hotels

a) Ornamental flowering trees

- *Bauhinia purpurea*
- *B. variegata*
- *Brownea coccinia*
- *Callistemon lanceolatus*
- *Cassia javanica*
- *C. nodosa*

- *Cassia fistula*
- *Jacaranda mimosaefolia*
- *Saraca indica*
- *Peltophorum ferrugineum*

b) Ornamental foliage trees

- *Polyalthia longifolia*
- *Araucaria excelsa*
- *Ficus religiosa*
- *Ficus benjamina*
- *Melia azadirachta*

6. Religious places

a) Trees for planting in temples (flowering and scented)

- *Cassia nodosa*
- *Citharexylum spinosum*
- *Magnolia grandiflora*
- *Michelia champaca*
- *Millingtonia hortensis*
- *Mimusops elengi*
- *Plumeria rubra*
- *P. alba*
- *Saraca indica*

b) Trees for planting in mosques (shade trees)

- *Peltophorum ferrugineum*
- *Polyalthia longifolia*
- *Pongamia pinnata*
- *Putranjiva roxburghii*
- *Samanea saman*
- *Thespesia populnea*

c) Trees for planting in churches (for calmness and concentration)

- *Amherstia nobilis*
- *Araucaria excelsa*
- *Callistemon lanceolatus*
- *Cassia nodosa*
- *Ficus lyrata*
- *Polyalthia longifolia*

7. Home garden

a) Trees for providing shade in houses, terraces & outdoor living areas

- *Callistemon lanceolatus*
- *Cassia nodosa*
- *C. javanica*
- *C. renigera*
- *Grevillea robusta*
- *Jacaranda mimosaefolia*
- *Lagerstroemia flos-reginae*
- *Phyllanthus emblica*

b) For small gardens or compounds

- *Bauhinia purpurea*
- *B. variegata*
- *Brownea coccinia*
- *Callistemon lanceolatus*
- *Cassia javanica*
- *C. nodosa*
- *Grevillea robusta*
- *Jacaranda mimosaefolia*
- *Mesua ferrea*
- *Spathodea companulata*
- *Saraca indica*

c) Trees for cut flowers

- *Amherstia nobilis*
- *Bauhinia variegata*
- *Cassia fistula*
- *Cassia nodosa*
- *Couroupita guianesis*
- *Delonix regia*
- *Jacaranda mimosaefolia*
- *Lagerstroemia speciosa*
- *Peltophorum ferrugineum*

d) For scented flowers

- *Alstonia scholaris*
- *Cananga odorata*
- *Couroupita guianensis*
- *Citharexylum spinosum*
- *Dillenia indica*
- *Magnolia grandiflora*
- *Michelia champaca*
- *Mimusops elengi*
- *Plumeria rubra*
- *P. alba*
- *P. acutifolia*

8. City forests and open green belts

- *Acacia auriculiformis*
- *Bauhinia variegata*
- *Butea monosperma*
- *Erythrina indica*
- *Kigelia pinnata*
- *Lagerstroemia flos-reginae*
- *Melia azadirachta*

- *Millingtonia hortensis*
- *Pongamia glabra*
- *Pterospermum acerifolium*
- *Samanea saman*
- *Thespesia populnea*

9. Trees for planting along banks of lakes and ponds

- *Casuarina equisetifolia*
- *Lagerstroemia flos-reginae*
- *Melia azadirach*
- *Terminalia arjuna*
- *Thespesia populnea*

10. Trees suitable for different soil and climatic conditions

a) For marshy areas

- *Casuarina equisetifolia*
- *Lagerstroemia flos-reginae*
- *Melia azadirach*

b) For dry localities

- *Acacia auriculiformis*
- *Butea monosperma*
- *Cassia fistula*
- *Erythrina indica*
- *Jacaranda mimosaefolia*
- *Kigelia pinnata*
- *Melia azadirachta*
- *Peltophorum ferugineum*
- *Plumeria alba*
- *P. rubra*
- *Pongamia pinnata*

c) For saline soils

- *Azadirachta indica*
- *Butea monosperma*
- *Eucalyptus citriodora*
- *Madhuka indica*
- *Phyllanthus emblica*

d) For moist and humid areas

- *Amherstia nobilis*
- *Bauhinia variegata*
- *B. purpurea*
- *Brownea grandiceps*
- *Cassia marginata*
- *Cassia nodosa*
- *Lagerstroemia speciosa*
- *Samanea saman*
- *Saraca indica*
- *Spathodea campanulata*

e) For acid soils

- *Butea monosperma*
- *Cassia fistula*
- *Casuarina equisetifolia*
- *Eucalyptus citriodora*
- *Melia azadirachta*

7

Planting, Care and Management of Ornamental Trees

Trees are perennial in nature and are valuable elements in a landscape. Trees provide shade, privacy and beauty but a tree in a wrong place may create problem in years to come. To avoid this, care should be taken while selecting trees for landscape. The ultimate size and shape of the mature tree, how much annual maintenance it requires, location and purpose are major considerations while selecting a tree. Matured tree height and tree diameter should fit it into the space when tree is planted. Shape of canopy, colour of flower and season of flowering are also to be considered.

Selection of site

"Landscape with poisonous element, abundance of stones, ant hills, holes, and gravel and having no accessibility to water is unfit for growing trees. Land which is even, has accessibility to water, and is covered with green trees is good for growing all kinds of trees" (Vrikshayurveda).

Above statements in Vrikshayurveda point out to the importance of site for planting trees. Different species of trees are available, suitable for planting in different soil conditions. An understanding of requirements of soil type, sunlight, temperature and water, of different species makes selection easier.

Soil plays a very important role in the life of trees. A light and well drained soil of mixed sand and clay is ideal for growth of trees. Water logged and low lying areas produce stunted growth. Leguminous trees flourish in poor sandy soils by fixing atmospheric nitrogen and making it available for growth. Then, there are trees which can cope with alkaline soil, such as *Butea frondosa*. There are trees which grow in dry rocky areas with minimum of soil, such as *Cochlospermum gossypium, Cassia fistula,* and *Plumeria sp.* These trees are ideal for covering arid hills.

Nursery tree selection

Healthy trees make our environment more pleasant and increase value of landscape. Selection of good quality planting material is very essential for raising healthy trees. Most of the trees are propagated by seeds. Seedlings for planting may be bare root plants or containerized plants. In case of bare root plants roots may be protected from drying out. Containerized plants are with a soil ball and they are more likely to survive than bare rooted plants. Select tree seedlings with branches all along the trunk and not just at top.

When purchasing bare-rooted tree seedlings, inspect the roots to ensure that they are moist and have numerous and healthy fine root hairs. Care should be taken to ensure the roots are kept moist in the period between purchase and planting. Prune broken or damaged roots but save as much of the root structure as possible.

Installing trees

Success and survival of tree in the landscape is determined by preparation of planting site, proper installation and good post transplant care. Each is critical in assuring that the plant survives its first year after transplanting.

Preparation of planting site

The place where tree is to be planted (the *site*) should be well drained. Soil should not be compacted or heavy with clay. If the soil contains too much clay, water will not pass through it quickly enough. It may be necessary to add sand and other organic material to loosen and lighten the soil. If the drainage is very poor, the site requires installation of drainage facilities. If the soil is excessively sandy, it may drain too quickly. In this case, addition of organic material may be necessary to increase soil's ability to hold water.

The hole which is dug should be at least 50% larger than root system or soil ball, it is receiving. If a bare root plant is being transplanted, hole should be prepared with a mounded bottom to allow roots to spread in a natural fushion. If a containerized plant is being used, hole should be flat bottomed.

Proper installation

Plant should be set into ground to same depth at which it was originally growing, or slightly higher if drainage is poor. With a container-grown plant, top of soil ball represents original ground level. With a bare rooted plant, there is usually a natural strain on bark marking former soil level. Duplicating this original soil level is very important because plant will die if placement is too deep or too shallow. Care should be taken to disturb root system as little as possible.

After the hole was backfilled with properly conditioned soil, soil is tamped very carefully with foot. This assures that there are no unseen air pockets below soil to collapse after planting. Gently raising and lowering plant helps to settle soil. A thorough, deep watering at time of transplanting is essential.

Many times, an application of mulch with a porous, coarse materials is made after planting seedlings. Mulching reduces evaporation and helps soil to maintain a more even temperature.

Deep and thorough watering of plant is necessary throughout the first growing season. Enough water should be used to wet the soil to a depth of 15 to 20 cm. Infrequent, deep watering is preferable to frequent shallow watering.

The American Forestry Association, in cooperation with National Urban Forest Council, developed new guidelines for properly planting a tree. Idea is to plant so that roots have a chance to grow into surrounding soil and produce healthy, vigorous branches, foliage, and roots.

New recommendations, based on recent research, suggest marking out a planting area five times the diameter of plant root ball and using a rotary tiller or shovel to loosen and mix the soil in the entire area to a depth of about 30 cm. Organic matter can be added to loosened soil as long as material is used uniformly throughout area.

In center of prepared area, dig a shallow hole to set the tree's root ball. Hole should allow root ball to sit on solid ground rather than loose soil. After the tree is properly situated, cut and remove the rope or wire basket holding the burlap and cut the excess burlap away from root ball. Position the tree perpendicular to ground with main stem growing straight up.

Backfill around root area and gently pack soil to remove major air pockets. Water can be used instead of your feet to help to settle soil around roots. This will also prevent excessive compaction. Upper surface should be level with or slightly higher than existing soil. Rake soil evenly over the entire area and cover mulch. Correctly planted trees can grow twice as fast and live twice as long as those improperly set out.

Staking and guy-wiring

Young trees should be able to support their own weight, but when they are transplanted, they often need time to reestablish themselves. When staking a tree, remember the following.

- Only stake the tree long enough for it to be able to stand on its own.
- Stakes should not be too tight - there should be room for tree to sway in wind.
- Stakes should not be too loose - the tree should not rub against stakes.
- Stakes should be buried at least 45 cm underground to provide ample support.
- Drive stakes into undisturbed ground before covering hole, to avoid damage to roots.

Larger trees should be guyed. To guy a tree, use three wires attached to three stakes. Thread a 30 cm piece of hose on each wire and have the hose-covered bit of wire loop around the tree at the lowest branch crotches. Make loops around the trunk very loose to avoid damaging the tree. Drive stakes into firm soil at least 45 cm deep outside the perimeter of the planting hole. Anchor the loose ends of wires securely to stakes. Various fabric straps and rigid staking systems are available that may be easier to use than hose and wire.

Stakes and guys should be removed before wire girdles the trunk. Generally, remove supports after one growing season for a 4 cm diameter tree seedling, two seasons for a 8 cm diameter tree seedling. For larger seedlings, guys may need to remain for three seasons or more. They should be inspected annually and adjusted to prevent trunk girdling.

Watering

Newly planted trees to survive their first year, have to get the right amount of water. Over watering is just as harmful as under watering. How often and how much to water depends upon the soil type and the amount of rainfall. On sandy soils, plants may require more water especially in summer while plants in poorly drained, clay soils need less frequent watering.

Fertilizing landscape trees

Maintenance programmes should be developed for trees in both residential and commercial landscapes. A good maintenance programme includes monitoring and controlling insect and disease problems, suppressing weed competition, and making timely applications of water, mulch, and fertilizer.

Tree fertilization is especially important in urban areas where soils are altered due to construction. These urban soils tend to be heavily compacted, poorly aerated, poorly drained, and low in organic matter. Even where soils have not been affected, fertilization may be needed as part of a maintenance programme to increase plant vigour or to improve root or top growth.

How and when to fertilize landscape trees depends on maintenance objectives (stimulate new or to maintain existing growth); age of tree (generally more for younger than for older plants) and plant stress levels. Visual inspection of trees is often the best overall factor to determine fertilizer requirement.

Poor or chlorotic leaf colour (pale green to yellow), reduced leaf size and retention, premature leaf drop, reduced twig and branch growth and overall reduced plant growth and vigour call for fertilizer application. In addition to observing signs of possible nutrient deficiencies on plants, soil and foliar analysis can be used to help determine or confirm whether supplemental fertilization is needed.

Nutrients can be applied in organic and inorganic forms. Slow release fertilizers and dry forms (granulated, pelletized, spikes, pulverized, encapsulated) are available now. Fertilizer may be selected considering type of plant, time of year, desired rate of plant reaction, application methods and equipment cost, proximity to water sources, effect of soil type and pH, type of deficiency, and results of a soil test or other sampling method. Most landscape plants benefit from a slow release nitrogen fertilizer that may be organic or inorganic.

Fertilizers can be applied either directly or indirectly to plants. When turf is fertilized, tree roots that extend into turf area absorb some of the fertilizers, and are therefore indirectly fertilized. Turf fertilization rates should be supplemented only if trees are showing symptoms of nutrient deficiency. Direct application of fertilizer involves incorporation into the backfill soil or placement in the planting hole at planting time. Compacted soil should first be aerated or raked and fertilizers are incorporated into the soil.

Fertilizer should not be concentrated around stem or trunk of a tree, but should be applied over plant's root zone. For trees and shrubs, fertilizer should be applied over an area twice as large as the crown spread or dripline. Since most landscape plant roots grow in top foot of soil, surface or shallow application is recommended. Fertilizer rates should be based on plant type, younger trees and shrubs generally receiving higher rates of N than mature plants. If a complete fertilizer is used, the ratio of N-P-K should be 3-1-2 or 3-1-1. In general, use 0.5 to 3 kg / 100 sq.m of root zone. Evergreen trees need less (0.5-1.5 kg) while deciduous trees commonly need more (1.5-3 kg). Reduce the rate when plants are growing in restricted areas (sidewalk cuts, parking lot islands etc.) or where roots of multiple plants overlap.

Factors affecting fertilizer uptake

Numerous factors affect rate of fertilizer uptake by trees. Following are the most important factors which affect fertilizer uptake.

- Fertilizer form (inorganic, fast release, or liquid forms are absorbed faster than organic, slow-release, or dry forms)
- Soil type (clay particles and organic matter adsorb or bind more nutrients than sand, so fertilizer application should be done more frequent in sandy soils, but with lower rates each time due to leaching potential)
- Soil moisture content and soil temperature (nutrient uptake is faster in moist warm soils)
- Plant vigour (plants under stress are less able to take up available nutrients due to damaged or reduced root systems)

Fertilizer should be applied when plants need it, when it will be the most effective, and when plants can readily take it up. A split application may be more beneficial. If water is unavailable, do not apply fertilizers as plants will be unable to absorb nutrients. During a dry season, fertigation (application of fertilizer through an irrigation system) can be beneficial.

Pruning ornamental trees

Correct pruning is an essential maintenance practice for trees in landscape. Pruning is the removal of a portion of a plant for better shape or more fruitful growth. It is easily done, but not so easily done correctly. Each time that a bud or branch is removed from a plant, it creates both a short term and long term effect upon the plant. The short term effect is how the plant looks immediately after pruning, and perhaps through the remaining current growing season. The long term effect is how plant appears after several seasons of growth without the part that has been pruned.

Trees should be examined annually for pruning requirements. Too often, pruning is ignored for several years. Then some trees become overgrown and often weak, making drastic pruning a necessity to bring the plant back to usefulness. Regular pruning will help to keep the plant in bounds and keep its growth vigorous.

Parts of a tree

Before beginning to prune a plant, it is necessary to have a basic understanding of the anatomy of plant.

In a tree, the lead branch is the most important branch. It is dominant over other branches, called scaffold branches. The lead branch usually cannot be removed without losing the distinctive shape of the tree. This is especially true in young trees.

The scaffold branches create the canopy of a tree. The amount of shade cast by the canopy of a tree is directly related to number of scaffold branches and size of leaves. When it becomes necessary to remove a branch from a tree, removal usually occurs at a crotch, that is, the point at which a branch meets the trunk of the tree or another larger branch. It is always desirable to leave the strongest branches and remove the weakest. Where crotch union is wide (approaching a right angle), branch is strong. Where the crotch union is narrow, branch is weak and may break in a heavy wind.

Two other types of branches often found on trees are suckers and water sprouts. Suckers originate from underground root stem. Water sprouts develop along the trunk of the tree. Neither is desirable for a healthy tree and both should be removed.

Reasons to prune

Pruning is necessary to maintain a healthy and vigorous tree. Specifically, pruning is practiced for following the purposes.

Maintain or reduce plant size

Pruning prevents a plant from overgrowing its space in the landscape and eliminates the need for drastic cutting of crowded, overgrown plants. It allows for growth of plants under or adjacent to pruned plant. It reduces leaf area on newly planted trees. This promotes survival through transplanting and consequent root loss.

Remove undesirable growth

Pruning encourages plant vigour through removal of weak, overcrowded growth. Such thinning often improves visual balance or symmetry of plant.

Remove dead, diseased, or broken branches

Pruning aids in maintaining shape, vigour, and health of the plant by removing dead and diseased branches.

Stimulate flowering and fruiting

Removal of current year's old, faded flowers and fruit clusters promotes flower buds.

Rejuvenate and restore old plants to vigorous growth

Proper pruning restores a youthful, natural growth habit in certain overgrown trees.

Prevent damage to life and property

Pruning minimizes hazard of limbs interfering with power lines or overgrowing structures. It also removes weak crotches before limbs break in strong winds and opens blocked sight lines caused by overhanging limbs at driveways or street corners.

Shape plants in an artificial form

Pruning and shearing can be used to shape trees as hedges or for rigidly formal espaliers or topiaries.

Time of pruning

Timing of pruning is based on flowering, fruiting, or growth habits of a plant, its tendency to "bleed," and the fact that pruning usually stimulates a flush of re-growth. Most plants can be pruned at almost any time of year without jeopardizing basic survival. However, it is preferable to prune specific plants at specific points in the year. The best time to prune is when the plant recovers the fastest. Damaged plants should be pruned soon after the damage occurs.

Tools for pruning

Proper pruning requires proper tools, employing correct methods, and applying methods accurately to appropriate plant. Tools should be of good quality and they should be sharp. Keep them in good condition by lubricating regularly, cleaning to prevent rust, and only using them for their intended function. Some of the pruning tools are hand pruner, lopping shears and pruning saw.

Basic procedures in pruning

Basic techniques of pruning are similar to all situations of pruning.

Pinching is usually done by hand, and this is a good way to control plant size. *Thinning* completely removes some branches back to a main branch, trunk, or soil line. *Heading back* involves shortening branches back to a good bud or lateral branch. A proper heading back cut should not leave a stub. Make your cut about 1-2 cm above an active bud or lateral branch. *Jump cutting* allows scaffold limb to be removed without stripping off a long slice of bark with it as it falls. The jump cut requires three cuts for safe removal of a limb. Final cut should remove stub of limb as close to the trunk as possible.

Wound dressing or tree paint is not essential; recent research has pointed out that tree paint or wound dressing is not as advantageous as previously thought. Dressings may actually harbor disease organisms rather than exclude them. It has also been determined that wound dressing slows wound callusing process, (often called healing) rather than speeding it up. The only reason for painting a pruning cut is a cosmetic one. When appearance is a factor, painting should be done with latex paint. A good, clean unpainted pruning cut, while perhaps unsightly at first, will probably callus faster than a painted one. On the basis of tree health alone, pruning cuts should not be painted.

Basic safety rules for pruning are the following.

- Call a professional for large trees or for jobs you don't have the equipment for.
- Keep all equipment sharp and in good repair.
- Use equipment only for the job it was designed to do.
- Be conscious of electric lines when pruning near them.
- If a power line is touching a tree limb, call the power company fast and stay clear of the tree.
- Never climb a tree without a safety rope, with or without a ladder.
- Keep your fingers clear when using hand clippers.
- Use care in handling pruned limbs and brush to avoid eye injury.

Procedures for pruning trees

Actual limbs and branches that are removed from a tree are determined by reason for pruning. If the objective is to remove diseased portion, the cut should be made through healthy wood between trunk or crown and infected part. Cut should never be made through diseased wood or just behind it. If this is done, pruning tool becomes contaminated and may transmit disease causing organism to healthy parts that may be pruned later.

If the pruning is being done for overall health and appearance of the tree, those branches, which are growing into the centre of the plant, are removed. Limbs and twigs, which are growing across other branches, may crowd and harm the plant. Such branches should be selected for removal. If more than one limb originate at a tree crotch, the strongest should be left and the others removed.

If the pruning is done to reduce size of the plant without altering its natural shape, careful selection of both pruned branches and those allowed to remain must be made. Major structural limbs and twigs must be left so that no holes appear in the plant. The fact that many secondary branches stem from one older branch is often overlooked. Result is that removal of older branch creates a tree with an entire side missing.

If the purpose of the pruning is to create denser foliage, it is the centre shoot, which is shortened or removed. This encourages lateral buds to grow and create two shoots, where formerly there had been only one. Result is the added plant fullness.

On most shade trees, top-most growing point is critical in achieving a form typical of species. This growing point is called the central leader, and there should be only one. It should not be cut to retain characteristic form of the tree.

Low-hanging branches cause injury to individuals mowing lawn or walking on the street. Also, branches sometimes rub against the house and roof. These branches should be removed by jump cut method.

Trees should not be topped as far as possible. Topping is the senseless practice of indiscriminately removing a majority of a tree's branches. Topping violates most commonly accepted methods of proper pruning and it is an assault on health and beauty of trees. There are five important facts to remember about tree topping

- Topping will not make trees safe; it actually creates hazardous trees.
- Topping makes a tree more susceptible to storm damage.
- Topping makes a tree more prone to insect and disease problems.
- Topping is a dangerous, expensive and ugly mutilation of a tree
- Topping is a waste of money.

Newly transplanted trees

During transplanting, bare root trees suffer root damage as well as a loss of roots. To compensate for this loss of roots, leaf area of the plant should be reduced by 1/3 to reduce water loss due to transpiration and evaporation. Thinning should preserve natural form of newly planted trees. Remove branches and parts of branches by cutting at different lengths.

When the tree is approximately 2 years old, establish good branch spacing. Branches should arise alternately from main stem and be well-spaced. This allows stronger limbs to develop. Select branches with the widest angles in the crotch. Wider the angle, stronger is the limb attachment.

Transplanting

When a tree is transplanted, it usually passes through a period of shock. Nature provides a balance between root system and leaf crown. If the tree came from a reputable nursery, roots were probably pruned for several years to form a compact ball of roots that could be taken up with the tree. Such pruning avoides undue disturbance of this natural balance. But, if a tree is dug from woods or field, where no previous root pruning has been done, its chances of survival are not too good. If this is done, cut out at least one-third of branches, at transplanting time. By doing this, number of leaves that will grow during following season will be reduced, which increase chances for survival. Another way of reducing water loss is to spray the plant with an antitranspirant. An antitranspirant is a liquid, which forms a plastic like seal over leaves, temporarily reducing their water loss.

Wrapping tree trunk with a tree wrap of burlap stripping or heavy paper can also help to reduce water loss and prevent sun scald (drying and blistering of bark caused by winter sunlight). The wrap should either be removed or allowed to decompose in no more than a year and a half. Before wrapping the tree, be sure to remove any labels, cord, or wire from branches or trunk. Allowing them to remain, could cause serious injury as tree grows.

Tree problems

If tree leaves have spots caused by fungi, bacteria or viruses; or affected by pests, the trouble can be identified and recommendations given. But, often leaves are dead or dying merely from lack of water due to disturbance elsewhere in the tree. These leaves give little or no clue to direct cause of death. Often the fungi present on specimens are not the direct cause and start to grow after the twig or branch dies. In many cases, answer to a tree problem can be determined only after a careful inspection of tree and its surroundings. Here are some clues that may help in solving a tree problem.

Bulldozer blight

This "blight" is common in recent development areas where wooded sections of land have been cleared and buildings are constructed among remaining trees. The trunk may be so badly damaged that little bark remains. Other trees may have been protected so that earth-moving equipment did not damage the trunk. However, heavy machinery might have compacted the soil and injured roots.

Grade change

Changing grade under a tree means adding or removing soil in area of root system. Removal of even a few inches of soil will take with it some of the feeder roots and cause shock to tree. Injury produced by lowering grade is negligible, however, compared with that caused by adding soil to raise grade.

Drought

Trees growing in poor locations may have a hard time getting enough water to supply their needs even in normal years. During an extreme drought, moisture in soil may be so reduced that many of the feeding roots will die. Leaf scorch may result during even a short period of dry weather.

Excessive moisture or poor drainage

In areas where subsoil is heavy clay, a recently transplanted tree may die, because of excessive water in soil. Filling the planting hole with good sandy loam or peat does not change the water problem. If excess water cannot drain out of planting hole, the newly planted tree may not have a good chance for survival.

Dig a hole where the tree is expected to be planted and pour in several buckets of water. If much water is still there the next day, then expect trouble with a tree planted there. In these areas, plant the tree "high" in the hole rather than setting it too deep. It may be well to put drainage tile in bottom of hole to drain off excess water.

Trees surrounded by pavements or buildings

Trees growing with a street on one side and a sidewalk on the other are usually the first to show results of drought or high temperatures. Allowing water to run slowly from a hose in any open soil under crown spread will often do much to help the tree. A partial thinning of top of tree will help to balance root and leaf ratio and allow rain to penetrate the crown to soil around the trunk.

Girdling of roots

When planting a tree, in many cases, hole is dug too small to accommodate outspread roots. Also, roots may be wound around base of the tree so that all fit into undersized hole. Meanwhile, trunk grows in diameter until roots are in direct contact with it. A few years later, roots, which are now even larger, will be cutting into the tree and choking it much as would a rope around neck of a man. Such "girdling roots" can cause untold damage to growing shade trees.

The first symptom of a girdling root is the gradual decline of a branch or branches on one side of the crown. Every year, more and more branches will die, until finally one whole side of the tree may be dead. Before this stage is reached, an examination of trunk at the soil line will reveal presence of girdling roots, regardless of whether they are at or below the surface of the soil. Most trees, even while still relatively young, will show a slight to moderate buttressing of trunk at soil line. Troublesome root or roots may be cut from trunk with a sharp chisel.

With increase of container-grown stock in recent years, many trees start out with root system badly disfigured. Roots of seedlings in containers quickly reach confines of the container and then start to encircle soil ball. In later years, small encircled roots enlarge and result in strangulation of young trees.

Lightning damage

Although a stroke of lightning that kills a tree will usually leave loose bark hanging from top towards ground, a tree may show no outward signs of stroke. Some trees may be killed completely. On others, a few of the topmost branches will be killed. On some trees, when trunk has been killed from top to bottom, not only will the leaves remain green through season, but also the tree will again produce leaves in next season. If bark around the circumference of trunk has been killed, the tree can live for a year or two.

Septic tank injury

Strong chemical agents used in the household disposal system may escape from the septic tank into the surrounding soil and cause injury to trees. Occasionally, leaks or breaks in septic tank grease traps buried in lawn areas, will injure tree roots.

Spray injury

Some spray materials that are non injurious to certain species of trees, for instance, are extremely toxic to other species, or perhaps the concentration used on one type of tree is not adapted to another.

Herbicide or weedicide injury

Weedicide injury could be suspected if leaves are twisted, cupped and unusually small with light-coloured veins in contrast to darker green interveinal portions. A weedicide, applied to one corner of a yard or several blocks to 1.25 km away, may drift into a tree on a windy day and result in damage. At other times, these chemicals may be used in excess or too close to a tree, in which case, tree roots may take up some of the chemical and leaf injury results. Do not apply weedicides or herbicides under trees, near shrubs, or hedges.

Air pollution damage

Air pollution from industry, automobile exhausts, fires of all types, and fumes given off by chemical processes result in damage to trees and plants. In some cases, exact pollutant can be identified by merely recognizing characteristic injury symptoms it produces on some plants. In other cases, a pollutant may be suspected after observing localized area in which plants are damaged.

Other causes for sick trees

- Wires or chains used for clothelines or fencing left on a tree may be overgrown by the bark as the tree enlarges. In later years, these metal constrictions can cut off the sap stream and kill the tree years after the wires or chains are overgrown and no longer visible.
- Young or recently transplanted trees are often girdled at the soil line by mice and rabbits. Examine the bark at and slightly below soil line for chewing injuries.
- Lawn mowers banged against base of a tree are notorious for causing injury.
- Some tree species that are not native to the locality may be killed by low winter temperatures or they may not be able to endure extreme summer temperatures.
- The Southern sides of trees may have a dead streak extending from the soil line well up the trunk of the tree. This is usually the result of winter injury resulting in death of cambium or growing layer just beneath the bark. The extremes in day and night temperatures, during winter, on the southern side of a tree trunk is enough to cause cambium injury in some trees.
- Street trees are occasionally injured by salt used during winter to remove ice from pavements in colder countries. Roots are killed when the ice thaws and water carries salt to area of tree roots.
- Lawn parties, during which brine is spilled from ice cream freezers, may result in injury to small trees.
- Some chemicals, such as pentachlorophenol used to prevent rotting of commercial timber, will injure plant materials. Leaves of trees hanging near a treated fence or outdoor furniture may turn brown.
- Finally, some trees, like some human beings, just won't do well in a neighborhood - for no particular reason.

Here are six easy ways to have the best looking trees

- Plant the right tree in the right place.
- Water and mulch newly planted trees.
- Protect trees from mowers and other equipments. Avoid growing grass right up to base of tree.
- The soil should be firm around trunk, but don't pack the soil tightly when planting.
- Prune according to accepted methods.
- If you think a tree needs to be topped, better cut off the tree down and plant a smaller variety.

8
Description of Trees for Landscaping

Ornamental trees

Scientific name : *Acacia auriculiformis* syn. *A. moniliformis*
Common name : Golden shower
Family : *Fabaceae*

Moniliformis means 'like a string of beads' signifying the arrangement of flowers on the spike. The tree is native of tropical Australia. It is valued for afforestation and beautification. It grows in warm climate under both dry and moist atmosphere. It is a medium-sized, thornless, quick growing, evergreen tree of 13 to 15 m height.

Leaves: True leaves are present only in early seedling stage. The apparent leaves are the modified midrib called phyllode, leathery, sickle-shaped, tapered at each end, up to 18 cm long and 6 cm broad; nerves parallel.

Flowers: Yellow, scented, sessile, in dense spike near the end of branches. It grows rapidly and comes to flowering on the third year. Profuse flowering occurs in September-October; occasional blooms are observed between March and December.

Fruits: Pods 1.2 cm broad, margin wavy, twisted in the form of a spring containing black seeds.

Propagation: Propagated from seeds

Landscaping uses: It is grown as an avenue tree used in large scale for tree planting programme due to the capacity for rapid biomass production.

Scientific name : *Acacia mangium*
Common name : Mangium, brown salwood
Family : *Fabaceae*

Indigenous to North-eastern Australia, Papua New Guinea and eastern Indonesia.

The tree may grow upto 20 m, with straight bowl, usually grows up to 15 m in height; bark dark brown to brown, thick, rough, furrowed longitudinally.

Leaves: Phyllodes large, upto about 25 cm long.

Flowers: Small, white, cream or light yellow, in axillary spike, upto 10 cm long; pods initially straight and green, later twisted and blackish brown when ripe.

Scientific name : *Adenanthera pavonina*
Common name : Red sandalwood, Bead tree, Red wood, Peacock tree
Family : *Fabaceae*

Pavonina refers to Latine word 'peacock'. A native of China and Malaysia. A medium to large, 10 to 16 m tall, quick-growing tree with rough, dark grey bark.

Leaves: Bipinnate, dark green, divided into large number of elliptic oblong, blunt leaflets, 2 to 3 cm long, arranged on either side of the midrib.

Flowers: Minute (4 mm long), yellowish and scented, arranged on spikes (5 to 15 cm long and 1.5 cm broad) springing singly from near the base of the leaves and grouped in open cluster near the end of the branches; calyx campanulate and the teeth short; the base of 5 minute petals spreading in the form of a star; 10 prominent free stamens with anthers set on the tip. Flowering season April - September

Fruits: Long and narrow pods (15 to 22 cm long), leathery valves twisted and bend as they open; seeds deep red and hard.

Propagation: The tree is propagated from seeds which do not germinate easily unless kept moist for some days before sowing. It can also be raised from large cuttings planted during the monsoon.

Landscaping uses: This tree is recommended for planting along the roads and in parks and large gardens.

Scientific name : *Albizia lebbek*
Common name : Vaka
Family : *Fabaceae*

Lebbek is an Arabic name. The tree is a native of tropical Asia and Africa. It is found in the warmer parts of India, Sri Lanka, Myanmar, Malaysia and China, and is cultivated in various parts of America. A large deciduous tree (10-20 m) with spreading branches, brownish grey bark with numerous short cracks.

Leaves: Bipinnate; main rachis pubescent or glabrous with a large gland on the base of petiole and pinnate; 2-4 pairs, 10-15 cm long; leaflets 2.5-4 cm long and 1.5-2 cm broad, pale green, blunt, unequal sided, 5-9 pairs on each pinna with glands between their bases. Leaves fall at the end of cold weather when dry straw-coloured fruits are seen even from a long distance.

Flowers: Flower-heads globose, umbellate, containing many flowers, creamy white, fragrant and clustered 2-4 together on stout axillary peduncles, 4 cm long to the end of prominent stamens; calyx small with triangular teeth; corolla tube (2 cm long) numerous greenish white stamens, 3.5-4 cm long, united at the base into a short tube, extended far beyond the other parts of the flower to form the conspicuous part of the bloom; pods 10-20 cm long, 2-4 cm broad, linear oblong, pointed, pale straw coloured, papery, markedly veined, smooth, and shining when ripe and dry.

Fruits: Seeds 4-12, ellipsoid-oblong, flat, depressions on the faces, pale brown.

Propagation: It is propagated from seeds.

Landscaping uses: The tree is grown in parks and gardens for its beautiful crown and sweet-scented flowers. It has a shallow root system and is, therefore, susceptible to strong winds, particularly during the rains when the soil is wet. It is short lived and attacked heavily by bark-eating caterpillars.

Scientific name : *Amherstia nobilis*

Common name : Tree of Heaven, Queen of the flowering trees

Family : *Fabaceae*

Amherstia nobilis is a native of Burma, it is one of the most beautiful trees. It is a medium-sized, evergreen tree attaining a height of about 8-12 m. Leaves are compound, paripinnate and drooping , young leaves with pale copper or purple colour which turns green when mature. Flowers are large, sprays of vermillion red tipped with golden yellow spots on petals and arranged in long, pendent candelabra like racemes arising in the axils of the leaves. The tree flowers in November and April. The flowers last only for two to three days. Fruit is a flat pod.

Leaves: 30-45 cm long, divided into 6-9 opposite pairs of narrow, pointed, smooth leaflets (12-20 cm long by 1.5-2.5 cm wide) set on either side of a central midrib; young leaves which at first hang limply from the ends of the branches and in this flaccid state assume various beautiful shade of pink, copper, mauve or purple, soon turn bright shining green on becoming firm. New leaves are produced several times a year.

Flowers: Arranged in long (50-75 cm), pendant, candelabra like clusters arising from the axil of the leaves, hang from all parts of the tree; upper flowers, usually having longer stalks than the lower ones, give the clusters a conical shape; petals brilliant red, edged with yellow, upper petal the largest (5 cm long). Flowers appear from February to April.

Fruits: Pods flat, hairless and smooth, beaked, 8-12 cm long and 3 cm wide, containing 4-6 seeds. Young pods brilliant crimson with greenish markings.

Propagation: Fruit set is often uncertain and propagation is done mostly by air-layering.

Landscaping uses: The tree is not successful in dry and arid regions or where there is danger of frost. It thrives in a porous soil rich in organic matter, moist climate, in areas of good, well-distributed rainfall, but it does not seem to thrive near the sea. Young plant may die unless they receive meticulous attention. The tree is planted in gardens for its beautiful crown, leaves and flowers.

Scientific name : *Anthocephalus chinensis* syn. *A. indicus*
Common name : Kadamba
Family : *Rubiaceae*

Chinensis and *indicus* refer to the countries of origin. In India *kadamba* has a mythological association with Lord Krishna. About 2000 years ago vast kadamba forests reportedly existed in the Vrindaban region, between Mathura and Bharatpur, remnants of which can still be seen. A tall deciduous tree (15-20 m) having a perfectly straight trunk, fairly smooth bark, marked with numerous longitudinal cracks, peeling off in small rectangular scales; branches horizontal with a tendency to droop and form a dense rounded crown of foliage. It grows best in warm and humid climate and rich loamy soil. The growth is rapid in the first 6-8 years and the tree attains maximum size in about 20 years.

Leaves: 12-22 cm long, much larger in young plants, elliptic-oblong or ovate, acute or shortly acuminate, dark glossy green above with paler veins, lighter undersurface and usually covered with fine hairs down.

Flowers: Numerous, small, scented, borne on flower-heads, solitary, globose, 4-5 cm in diameter, on 2.5-4 cm long peduncle at the end of the twig; calyx lobes oblong and persistent, pale greenish and cream-coloured; 5 petals partially jointed to form a small funnel-shaped tube, glabrous and orange in colour and as they grow outwards beyond the calyces the heads become compact orange spheres, 5-8 cm in diameter. Flowers appear from June to August. The tree comes to flowering when 4-5 years old.

Fruits: Composite, consisting of a fleshy globular mass on which are inserted the fruits of individual flower separately; on ripening they turn yellowish or brownish and attain 5-6 cm in diameter; seeds small and numerous.

Propagation: The plant is propagated from seed.

Landscaping uses: This beautiful tree is admired primarily for its golden ball of flowers with delicate scent and secondarily, for its mythological connections. It is often planted near houses and on roadsides as a shade tree, and is very suitable for this purpose owing to the dense shade given by its handsome, large foliage, its straight trunk and stately outline.

Scientific name : *Butea monosperma syn. B. frondosa*
Common name : Flame of the forest, Parrot tree
Family : *Proteaceae*

Monosperma in Greek means 'with one seed' and *frondosa* in Latin meaning 'covered with leaf'. A medium sized (10-12 m high) deciduous tree, untidy in growth and rugged in shape; trunk crooked and bark thick, fibrous and light or grey in colour; young shoots downy tomentose. The tree is frost-hardy, resistant to drought and tolerant to salinity to some extent.

Leaves: trifoliate; leaflet leathery, finely silky below when young, turning glabrous with age, nerves conspicuous; terminal leaflet obovate, 15-18 cm long, 14-15 cm broad, hard, leathery; leaves fall off during winter.

Flowers: large, about 5 cm long on racemes, upto 15 cm long; pedicel brown, velvety; calyx 1.2 cm long, velvety brown externally, silky white within; petal colour variable, standard being orange, outside grey, silky, pubescent, lateral petals similar to the standard, but narrower, the keel petals united to form a beak; stamens 10, 9 joined, one free. When in flower the tree looks as if covered with flame.

Fruits: Pods profuse, 10-15 cm long, 2-4 cm broad, pendulous, strongly nerved, pale green, one seeded, covered with silvery hairs and give the impression of foliage, yellowish grey when ripe, flower colour may be yellow, orange, orange-scarlet and in variety of shades. Flowering starts in February, continues for about a month and fruits ripen in April.

Propagation: It is propagated from fresh seed which germinates easily. The young seedlings produce bulb-like swelling at the base of the stem and may die back during hot weather, new shoots developed from these bulbs at the advent of the rainy season.

Landscaping uses: The tree can be used for planting along medium sized roads and also in parks and gardens, specially in group.

Scientific name : *Caesalpinia coriaria*
Common name : American sumac, Divi Divi
Family : *Fabaceae*

Coriaria in Latin meaning 'used for tanning'. It is a native of Central America and West Indies. It is mainly cultivated in tropical countries. A small, deciduous, umbrella-shaped, spreading tree, 8-10 m high with drooping branches. The tree grows well in dry climate.

Leaves: bipinnate, upto 22 cm long and 12 cm broad with 13-17 pinnae, each 5-8 cm long; leaflets 40-60 on each pinna, 8 mm long and 2 mm broad, oblong, dark green above and greyish below.

Flowers: Scented flowers (4-5 mm across) appear in dense terminal panicles near the end of the branches; corolla 5 mm long and greenish yellow; filaments reddish; pods 3-4 cm long, 1 cm broad, twisted spirally, smooth and pale to blackish brown in colour.

Landscaping uses: This is used as a shade tree in small gardens. It can also be used for planting in small avenues.

Scientific name : *Cananga odorata*
Common name : Ylang-ylang
Family : *Annonaceae*

Odoratum in Latin means 'sweet-smelling'. Tree is a native of Myanmar to Malaysia. It is cultivated in Java, Philippines, India, Sri Lanka and other countries for economic purpose and as an ornamental tree. An evergreen, quick-growing tree, 15-20 m tall, with ashy smooth bark, a straight trunk producing horizontal branches arching down at ends. Tree prefers a humid climate.

Leaves: Alternate, oblong or elliptic, often oblique, long acuminate with sharp point, 15-20 cm long, glossy above and slightly pubescent beneath.

Flowers: Many, pale yellow, very fragrant, drooping, on long pedicels, in 2- to 6 flowered, shortly pedunculate racemes; sepals 3, ovate, pubescent; petals 6, narrow, linear, equal, 6-8 cm long; stamens many, crowded. Seedlings come to flower in two years. Period of flowering is from May to October.

Fruits: In bunchs of 12 black berries, each one containing several seeds.

Propagation: It is propagated easily by seed. For commercial planting, 5 x 5 cm spacing is allowed and height of tree is maintained at 3-5 m.

Landscaping uses: Tree is well suited for parks and large gardens, particularly along boundary where they may be planted in lines. Being a soft wooded tree, it is better not to be planted along roadside. However, large scale planting of *C. odorata* is for manufacture by distillation of popular scent known as 'ylang ylang' and an oil known as cananga oil in Philippines and Indonesia.

Scientific name : *Cassia fistula*

Common name : Indian laburnum, Golden shower, Amaltas

Family : *Fabaceae*

Fistula in Latin meaning 'a pipe', which refers to shape of pods. Native of India, ascending upto 1000 m in foothills of Himalayas, also found in Myanmar, Thailand, Indonesia, Vietnam and Philippines. It is grown in tropical Africa, South America, West Indies and in all tropical and mid subtropical regions throughout the world for beautiful flowers.

A small to medium- sized, open- branched, deciduous and hardy tree of slow growth, reaching a height of 8-10 m; bark smooth, greenish grey on young trees, turning brown and rough with age. It is drought tolerant and, therefore, an excellent flowering tree for arid regions, though well - drained soil of moist locality is also suitable for growing this tree.

leaves: Large, upto 40 cm long, paripinnate; leaflets 8-16, opposite, glabrous when mature, ovate, acute or shortly acuminate, upto 12 cm long and 9 cm broad; new leaves often of a rich copper colour with a soft downy undersurface and remain pendulous and folded until fully grown. Tree becomes leafless in winter.

Flowers: Yellow, large, about 3 cm across, in long pendant racemes, 40-50 cm long; pedicels slender; calyx short, falling prematurely; corolla 5, large, showy, unequal, ovate, petals clawed; stamens 10, yellow, 3 with long curved filaments. Flowering occurs on bare branches in spring or early summer when it looks marvellous with numerous hanging bunches of bright yellow flower. In moist areas, flowers are often mixed with new foliage. Flowering is profuse but comparatively short in dry regions; in humid areas a few flowers may be observed even upto September.

Fruits: Pod, 30-60 cm long, indehiscent, green when young and black after ripening, with 40-100 seeds, embedded in a dark coloured sweetish pulp and separated by transverse partitions. After flowering is over, tree is decorated with long hanging sticks of green fruits which take nearly a year to ripen.

Landscaping uses: It is suitable along small and medium roads, also for planting in parks and gardens. The cattle repellent nature of the plant makes it eligible for open planting.

Scientific name : *Cassia nodosa*
Common name : Pink cassia, Pink mohur
Family : *Fabaceae*

Nodosa in Latin meaning 'knotty', in allusion to large knots seen on trunk and branches of this species. It is a native of hill tracts of Bangladesh,

Andamans, Malayan peninsula and Indonesia. It is widely grown in gardens of tropical countries throughout the world. A nearly evergreen tree of 12-16 m in height with knotted trunk, greyish or yellowish brown bark having narrow, deep horizontal clefts; umbrella-like crown sometimes touches ground towards periphery if grown in open space; young shoots silky.

Leaves: 20-30 cm long, paripinnate; leaflets 12-24, shining above, opaque beneath, elliptic-oblong, acute, 4-6 cm long.

Flowers: Bright pink or erect, pedunculate racemes rising prominently from nodes or leaf scars almost along the entire branching system; bracts narrow, lanceolate, pubescent, persistent; sepals 5, 5 mm long; petals 5, 2-2.5 cm long, narrow and pointed at both ends, bright pink, fading to almost white before falling, thus giving cluster a bicolour appearance. Flowering occurs during March-April along with new leaves and continues till July. Tree looks very pretty during flowering season when whole plant is decorated with cherry-like blossoms.

Fruits: Pods cylindrical, 30-45 cm long, black, containing a number of seeds divided by transverse partitions.

Propagation: By seeds, seeds should be rubbed against a rough surface to wear away thick seed-coat and then sown to get better and quicker germination.

Landscape uses: It prefers warm, humid climate and needs protection from hot, dry winds. It is a favourite tree in gardens and parks as well as along city roads. It can also be used as a specimen tree with its graceful leaves and umbrella-like crown.

Scientific name : *Cassia renigera*
Common name : Burmese pink cassia
Family : *Fabaceae*

A medium-sized deciduous tree reaching 8-10 m in height with a short trunk and a few upright branches bearing numerous slender, drooping branchlets; bark fairly smooth and brownish grey and covered with small corky excresences.

Leaves: Upto 35 cm long, bearing 8-20 pairs of oblong, obtuse, leaflets rounded at base, membranous, softly hairy, 3-5 cm long, 1-1.5 cm wide.

Flowers: Scented, pink, turning whitish with age, appearing in April-May in short erect clusters often with basal leaves from old leaf scars along, arching branches; petals elliptic-oblong, 2-2.4 cm long; flowering takes place when 4-5 years old.

Fruits: Pods indehiscent, round, almost black, 30-45 cm long, 2.3 cm wide, internally divided by transverse partitions, each chamber containing a seed.

Propagation: Cassia renigera is propagated by seeds. It is not a long lived plant; It thrives better in moist climate, though grows under dry conditions and comparatively poor soil.

Landscaping uses: It is a good tree for small gardens and can be used for planting along small rounds in parks and townships.

Scientific name : *Cassia siamea*
Common name : Kassod tree
Family : *Fabaceae*

Siamea means 'Pertaining to Siam'. Indigenous to Myanmnar, Malaysia, East Indies and the Southern most part of peninsular India. A medium-sized evergreen tree reaching about 12-14 m in height; crown rounded, becoming straggling with upright and dropping branches; bark nearly smooth, grey and marked with slight longitudinal fissures.

Leaves: 20-30 cm long; leaflets 12-20, elliptic-oblong, tipped with a minute sharp point, 4 cm long.

Flowers: Yellow, 2.5 cm in diameter, in large terminal pyramidal panicle often 50-60 cm long and mixed with pods; bracts linear, stiff, much shorter than pedicels; petals 5, almost equal, about 2 cm long, rounded in shape. Flowering occurs from June to January with a peak in autumn; stray flowers are also found during mid winter.

Fruits: A flat pod, ribbon like when young, brownish or purplish in colour when ripe, 15-25 cm long and 1.5 cm broad, containing a number of seeds.

Propagation: Tree grows luxuriantly under humid conditions but does not prefer water logging. It is propagated by seeds and grows vigorously both under sunny or semi-shady situation and reaches flowering stage in 3-4 years.

Landscaping uses: It can be used for avenue plantation, but due to brittleness, branches are liable to break down in wind and heavy rains.

Scientific name : *Citharexylum quadrangulare* syn. *C. spinosum*
Common name : Zither-wood
Family : *Verbenaceae*

Quadrangulare means 'square', in allusion to shape of young stems. A native of West Indies, this tree is often grown in gardens of humid tropics. A small, deciduous tree, 5-6 m high; bark light grey; straight trunk and wide-spreading, drooping branches; branchlets 4 angled.

Leaves: Opposite, elliptic oblong, nearly entire, about 16 cm long; flowers white, scented, drooping racemes, upto 20 m long; petals unite to form 5 mm long tube with 5 lobes. Leaves turn red before falling down

Flowers: Flowers appear in March-May and beginning of rain, or occasionally at other times.

Fruits: Small, fleshy, red when ripe.

Propagation: This tree is propagated by seed or large cuttings. Branches of old stumpy tree may be cut back with advantage to be replaced by vigorous young ones.

Landscaping uses: This small tree can be planted in small gardens or at back of shrubberies or in group in medium and large gardens and parks for delicate scent of its flowers.

Scientific name : *Couroupita guianensis*
Common name : Cannon ball tree
Family : *Lecythidaceae*

Guinanensis means 'of Guinana'. A native of Trinidad and tropical America, tree is grown in moist tropical and subtropical zones of world.

A soft-wooded, 15-20 m tall, deciduous tree with a straight, stout trunk and rough, brownish grey bark; young parts pubescent.

Leaves: Alternate, oblong-ovate or broad lanceolate, acute, subglabrous, about 15 cm long.

Flowers: About 8 cm diameter, in woody pendulous racemes 75 cm to 1 m long, springing from trunk and lower part of thick branches; calyx top-shaped with rounded lobes; petals 6, not equal, concave, rounded, about 5 cm long, outside a mixture of white, yellow and pink and inside deep pink or crimson. Peculiarity of flower lies in curious arrangement of stamens which are numerous, divided into two sets, infertile ones set around pistil and fertile ones united to form a serpent's hood-like structure over pistil, whitish, vertical, about 2 cm.

Fruits: Spherical, brown, about 15 cm in diameter containing large number of seeds; when ripe, pulp turns yellow with a foul, stinking smell.

Propagation: Couroupita guianensis grows well in warm humid climate. Tree is propagated from seeds which should be sown fresh. Sweet-smelling flowers are produced throughout warm season and fruit requires nearly a year to mature.

Landscaping uses: Trees are planted in parks, around temples and along wide city roads.

Cauroupita guianensis (Cannon ball tree) regularly responds to seasonal climatic changes. The phenomenon was observed repeatedly. At the onset of monsoon, this is the first tree to start shedding leaves. It also sheds its leaves with onset of winter. But this tree sheds all its leaves in two hours due to bursting of a chlorine cylinder and starts sprouting afresh within 24 hours. It is also used as a pollution indicator tree.

Scientific name : *Casuarina equisetifolia*

Common name : Beefwood, Casurina, Australian oak

Family : *Fabaceae*

Equisetifolia means leaves like an 'equisetum' or horse tail. The tree is a native of Australia, Malaysia and the Pacific islands. It is A tall, straight -stemmed tree reaching a height of 30 m.

Leaves: The ends of branches thickly set with numerous long, slender, usually drooping branchlets, mostly deciduous, performing the function of leaves; branchlets joined, 15-20 cm long, the internodes 4-6 mm long, 6-8 ribbed, from a distance the branchlets look like the needles of pine.

Flowers: Pinkish, inconspicuous; male flowers appear in spikes at the end of branchlets, at the base of which the female heads (cones) are born on comparatively thicker twigs; male spikes about 2 cm long; female cones, very short-stalked, globular, about 2 cm in diameter; ovary single-celled, with two ovules of which one develops into seeds; male and female flowers are usually borne on the same tree, sometimes on the same branch, in certain cases flowers of each sex are borne on separate plants.

Propagation: Propagated from seeds.

Landscaping uses: It is a good tree in parks and large gardens. Branches are brittle and hence not used on avenue tree. Casuarina is very suitable for mass planting along the sea shore and river banks where spacing should be close to check soil erosion. It is good for hedge and topiaries. Wood is used for propping.

Scientific name : *Delonix regia* syn. *Poinciana regia*
Common name : *Peacock flower, Flamboyant, Flame tree*
Family : *Fabaceae*

'Regia' in Latin meaning 'royal'. A native of Madagascar. *D. regia* has spread to South Florida, South California, Bermuda, Sri. Lanka, India, Bangladesh and other tropical and sub tropical countries. A large, deciduous, quick-growing, open-branched, spreading tree, reaching upto 15-20 m; bark slightly rough, greyish brown.

Leaves: Bipinnate, upto 60 cm long, on stout petiole and 10-20 pairs of pinnae set oppositely on rachis, each pinnae with 20-40 pairs of oblong leaflets, 1 cm long and 3 mm wide. Tree becomes leafless in spring and flowers appear in April.

Flowers: 7-10 cm across, arises in broad, erect clusters near to the end of leafless branches on new shoots; petals 5, about 5 cm long, orange-scarlet except the upper one which bears yellow or white stripes; stamens 10, red. Flowering continues upto July, sometimes even upto August. It is one of the finest flowering trees of tropics and subtropics and there is variation in flower colour ranging from scarlet to almost yellow.

Fruits: 40-60 cm long, 5-7 cm broad pod, hard and almost black when fully ripe; it splits open to disperse 20-40 oblong, hard seeds.

Propagation: Commonly propagated by seeds. Seedling grows very fast, reaching about 3 m in same year and producing flowers when about 3 years old. It can be propagated from large cuttings also.

Landscaping uses: It prefers warm humid climate, particularly along the sea coast but cannot stand waterlogging and frost, used for avenue planting, planting in parks in groups and as accent plant.

Scientific name : *Dillenia indica* syn. *D. speciosa*

Common name : Elephant apple

Family : *Dilleniaceae*

Indica means 'of India'. The tree is a native of India. An evergreen tree of moderate size, 6-8 m tall with straight, short trunk; smooth reddish bark peeling off in scales and spreading branches forming a dense, round, shady head; prefers a deep soil and humid atmosphere.

Leaves: Large, lanceolate, coriaceous, serrate, hard, 20-35 cm long, densely set near the end of the branches; petioles hairy, 3-5 cm long, stout

Flowers: Solitary, large, nearly 15 cm in diameter, appear among leaves; sepals 5, green, orbicular, persistent, thick and fleshy; petals 5, white, obovate. Flowers appear during June and July and the fruits mature in September-October. Flowers fall off quickly.

Fruits: Spherical, green, 8-12 cm in diameter, constituted mainly of the large, growing sepals, put firmly one over the other compactly; seeds small, numerous, reniform, hairy, long, embedded in an almost transparent, glutinous pulp.

Propagation: Easily propagated from fresh seed; cutting can be used for vegetative propagation.

Landscaping uses: The tree is suitable for planting in gardens and parks. It is a good avenue tree, particularly for moist soil and marshy areas.

Scientific name : *Erythrina indica* syn. *E. ovalifolia*

Common name : Dadap

Family : *Fabaceae*

A small, soft-wooded tree, upto 10 m high, with grey branches, armed with dark brown prickles arising from corky swelling of bark. Plant is indigenous to India, Sri Lanka, Malaysia and Myanmar.

Leaves: Trifoliate; leaflets elliptic or elliptic-oblong, upto 15 cm long

Flowers: Pea-shaped, deep red, 3-4 cm long, in racemes of 10-20 cm long, arising several at end of branchlets. Tree bears very attractive flowers in February-April.

Propagation: By seeds and hard wood cuttings.

Landscaping uses: Plant is used for group planting.

Scientific name : *Eucalyptus ficifolia*

Common name : Red flowering gum

Family : *Myrtaceae*

Native of South-West Australia. A handsome medium-sized tree, 12-16 m high with spreading crown; bark dark, furrowed, rough and persistent. Suitable for subtropical areas.

Leaves: ovate-lanceolate, 8-15 cm long; inflorescence in terminal corymbs, 15-18 cm across.

Flowers: Umbels with 3-7 flowers, on angular pedicels, upto 3.5 cm long; flowers red, 4 cm wide, on slender pedicels, upto 3 cm long. The tree flowers in April-May and again in August-September, depending on the climate.

Scientific name : *Ficus benjamina*

Common name : Java fig, Java willow, Willow fig

Family : *Moraceae*

Benjamina probably refers to the supposed relation of the tree to the source of a resin or benzoin. This tree grows wild in Malayan Archipelago, also found in the Andamans and different parts of India in the wet regions. *Ficus benjamina* has several forms showing variations in growth habit, size and shape of the leaves.

Large, graceful, broad-headed, evergreen tree reaching 15-20 m with spreading limbs and drooping branches that nearly touch the ground along periphery; a few aerial roots develop from thick branches; bark smooth, grey.

Leaves: Small, 5-8 cm long, shining green, leathery, ovate to ovate - elliptic; fruits or receptacles in sessile, axillary pairs, globular or ovoid, 1 cm diameter; male, female and gall flowers in the same receptacle. New leaves are produced from the early part of summer along with the figs.

Landscaping uses: This tree prefers humid climate and moist soil, and is propagated from seeds and large, thick or small and tip cuttings. Very good shade tree, suitable for planting along highways and large avenues, also in large parks, traffic islands and in the courtyards of temples and such other places where people assemble and need shelter under its shade. It is a good pot plant for indoor decorations.

Scientific name : *Ficus elastica*

Common name : Rubber tree, Indian caoutchoue tree, Assam rubber tree

Family : *Moraceae*

Elastica is derived from 'elastic' referring to the property of Indian rubber, which is manufactured from the latex of the tree. *Ficus elastica* grows wild in the Himalayas from Nepal eastward upto 1500 m, Assam, Khasi Hills and upper Myanmar.

A huge, handsome, glabrous, quick growing, spreading tree reaching above 30 m with equal spread when grown in open, sending down aerial roots from the branches; trunk supported by high, branching buttresses.

Leaves: Alternate, thick, leathery, shining green or variegated or coppery-pink, 12-30 cm long, nearly one-third or more broad, oblong or elliptic; petiole stout, 1-6 cm long; stipules nearly 15 cm long, pinkish yellow or reddish, caducous; receptacles in pairs on old axils, enclosed while young by hooded involucres. Fruit sessile, oblong, nearly 1 cm long; male, female and gall flowers within the same receptacles; monandrous.

Propagation: Seed propagation is feasible, but cutting and air-layering is generally practiced. Vegetatively propagated plants make very good house plants, variegated and coppery pink varieties are preferred. The tree prefers moist soil and humid climate, can tolerate shade.

Landscaping uses: Suitable for roadside planting, specially in hills and also for large gardens, parks and public places.

Scientific name : *Ficus krishnae*

Common name : Krishna's butter cup

Family : *Moraceae*

Small tree, upto 10 m high; with grey bark, aerial roots wiry, a few.

Leaves: Alternate, elliptic and form a cup on the dorsal surface at base, 15-20 cm long, rounded at apex, entire, cuneate at base; receptacle axillary, sessile, solitary or in pairs, yellow, about 1.2 cm in diameter.

Landscaping uses: Native to India, often planted in gardens and in pots for the unusual cup-shaped leaves.

Scientific name : *Holarrhena floribuda*

Common name : Pala, Ivory tree

Family : *Apocynaceae*

A medium-sized, handsome tree reaching about 10 m in height; branches not many, often horizontal.

Leaves: Opposite, ovate or ovate-elliptic, 6-10 cm long, 4-6 cm wide.

Flowers: Small, white, scented, 1 cm long, 0.8 cm a cross, many in axillary clusters on short peduncles. Numerous clusters of scented flowers appear in summer months.

Propagation: Propagated from seeds and by cuttings.

Landscaping uses: This well-shaped tree can be planted in small gardens and in groups in large gardens and also along roads.

Scientific name : *Inga dulcis*

Common name : Madras thorn, Manilla tamarind

Family : *Fabaceae*

Dulce in Latin meaning 'sweet', in allusion to the edible pulp of the pod.

A large, evergreen tree, upto 15 m tall, with grey bark; trunk often crooked; branchlets armed with short, sharp spines arising near the base of the leaves.

Leaves: Bipinnate, pinnae 2; leaflets 2, oblique, obovate-oblong, 2.5-4 cm long. New leaves are copper-coloured.

Flower: White, sessile, on small globose heads arranged in long, narrow terminal clusters, about 1.5 cm broad; pods linear, twisted, 10-12 cm long; seeds 6-8. Flowers appear from January-April

Fruits: Fruits become red when ripe.

Propagation: Propagated from seeds or cuttings.

Landscaping uses: The tree is common in dry and humid regions of India and many other tropical countries, grown along roads and boundaries of large gardens and farms. It is also a popular hedge plant and respond to pruning.

Scientific name : *Kigelia africana* syn. *K. pinnata*

Common name : Cucumber tree, Sausage tree

Family : *Bignoniaceae*

This remarkable tree is indigenous to Mozambique and several parts of tropical Africa.

A moderate-sized, evergreen, spreading tree, 12-15 m high with short trunk, distorted branches, covered with rough greyish-brown bark. It prefers dry, arid, tracts but grow in humid tropics under good soil conditions.

Leaves: Clustered near the end of branches in opposite pairs of whorls, imparipinnate; leaflets green, rough, leathery, sessile, elliptic-oblong or ovate, 8-15 cm long.

Flowers: Evil-smelling, on 0.8-1.5 cm long, rope-like pendulous racemes from branches; corolla gamopetalous, with cylindrical tube and 2-lipped limb, deep chocolate-red, about 12 cm across; the flower lip deflexed and 3-lobed and upper one 2-lobd, wrinkled and striped outside. Flowers appear in March-May emitting foul smell but attracts flies. Flowers open during night and fall down before the following noon.

Fruits: Hard, cylindrical, indehiscent, cucumber-like, 15-30 cm long and 6-10 cm in diameter, hanging singly or a few on cord-like long stalks; contains many seeds set in fibrous pulp.

Propagation: The tree is propagated from seed.

Landscaping uses: It is a good avenue tree, specially in the arid regions and can be planted in large gardens and parks for its shade and peculiar-looking fruits. In several parts of Africa this tree in considered to be sacred and grown there with reverential care.

Scientific name : *Lagerstroemia speciosa syn. L. flos-reginae*

Common name : Giant crape myrtle, Pride of India, Queen of flowers

Family : *Lythraceae*

In Latin '*speciosa*' mean 'beautiful' and *flos-reginae* meaning 'queen's flower'. Common in East and South India; wild in forests of Myanmar and Sri Lanka, particularly in moist localities, swampy lands and on river banks. A very showy, large, deciduous tree with globular crown, reaching 15-20 m in height; trunk short, both trunk and limbs have smooth, greyish cream coloured bark which peels off in irregular flakes during increase in girth. Tree prefers warm, humid climate and moist, deep soil. It can stand waterlogging very well; survives submergence for several days.

Leaves: From elliptic to lanceolate, mostly opposite on short, stout stalks, glabrous, bright green above and lighter below, 15-30 cm long and 6-9 cm wide. Leaves fall off in winter, new shoots emerge in spring along with terminal inflorescence.

Flowers: Wide open, 5-7 cm across on large panicles, 30-45 cm long at tip of branches; petals mostly 6, about 3 cm long, clawed, rounded and crumpled, spreading, mauve in colour. Flowering starts in 4-6 years. Main flowering period is from April to June; occasionally a second flush appears, specially in young plants. In this species, a type is also seen with large leaves, longer spikes and larger flowers, often termed as *L. speciosa* 'Magnifica'.

Fruits: A capsule, elliptic or spherical, 2-2.5 cm, woody; seeds light brown, angular, about 1 cm long.

Propagation: Propagated by seeds, growth in first year is slow, but faster afterwards. Training is needed to build a clean stem.

Landscaping uses: Very good for roadside, canal side, group planting along a large lawn, or around water gardens and ponds, used as accent plant, dry fruits used for decoration.

Scientific name : *Magnolia grandiflora*

Common name : Lily tree, Laurel magnolia

Family : *Magnoliaceae*

An evergreen, pyramidal tree with smooth, grey bark; reaches 15-20 m in cool, humid climate, but 6-8 m in warm and dry regions; twigs and buds rusty and hairy. The tree prefers cool, humid climate and seldom sets seed in warm regions;

Leaves: Simple, alternate, evergreen, obovate to oblong, 12-20 cm long, tapered to both ends, shiny above, rusty-pubescent beneath, leathery, thick and firm.

Flowers: Buds egg-shaped, borne singly at the tip of twigs and coated with silky-white hairs; cup-shaped when open, sweet-scented, white, 15-20 cm across; petals 6-12, ovate; ovary cone-like. Flowering occurs in the beginning of summer; number of flowers, size and fragrance are more in cooler climate.

Fruits: Ovoid, 8-10 cm long; seeds many, red, suspended by threadlike structure.

Propagation: It is propagated by cutting and layering.

Landscaping uses: It is a very good accent plant. It is valued for the graceful foliage and lotus like flowers.

Scientific name : *Melaleuca leucadendron*

Common name : White bottlebrush, Cajeput oil tree, Swamp tea tree

Family : *Myrtaceae*

The tree is a native of Australia and Malaysia. It is a slow growing tree and requires warm and humid condition; it can tolerate salinity to some extent. An evergreen tree with straight trunk, short, slender and pendulous branches and thick spongy whitish bark, reaching 10-12 m.

Leaves: scented, green, alternate, elliptic or narrow lanceolate, leathery, 5-12 cm long, tapering to a short petiole; nerves 3-7, longitudinal.

Flowers: Whitish, without stalk, in axillary spike, 8-15 cm long, prolonging to bear leaves; stamens numerous, prominent, united at base into 5 bundles; petals 5, minute. Flowers appear in profusion throughout the hot season and rains.

Fruits: Greenish, cylindrical with depressed top, small, containing many seeds.

Propagation: The tree is propagated by self-sown seeds than when sown in the nursery and also from cuttings.

Landscaping uses: It is a good tree used around water garden. It can be planted in the lawn to provide a good background for shrubs.

Scientific name : *Michelia champaca syn. M. aurantica*
Common name : Golden champa, Yellow champa
Family : *Magnoliaceae*

Champaca is Sanskrit name of the tree. Tree is indigenous to low hills in many parts of India and in Malaysia. An evergreen tree, large (25 m or more) at its native region and medium (12 to 15 m) in gardens of plains, with smooth, grey bark and cylindrical or conical crown;

Leaves: Alternate, simple, 15-25 cm long, 5-10 cm broad, lanceolate; leaf-buds covered with deciduous stipules.

Flowers: Yellow or dull orange, strongly scented; 5 cm across, perianth 9-15, outer oblong, inner linear. Flowering continuous throughout summer and rains at intervals with a few stray blooms all along.

Fruits: Fruiting spike compact, 8-15 cm long, containing many ovoid capsules, 2 cm diameter; seeds reddish, turning brown.

Propagation: Tree prefers moderate temperature, humid atmosphere and deep, rich soil. It is propagated easily from fresh seeds. Seedlings grow quickly and come to bear in 3-4 years.

Landscape uses: Trees are largely planted in gardens and can serve as an avenue tree in protected places. It cannot stand waterlogged condition even for a short period. Flower used for extraction of oil.

Scientific name : *Millingtonia hortensis*
Common name : Indian cork tree, Tree jasmine
Family : *Bignoniaceae*

'Hortensis' in Latin meaning 'pertaining to a garden'. A native of Myanmar, tree is widely grown in many tropical countries. A columnar, evergreen tree with straight trunk, reaching 20 m or more; bark thick, corky, yellowish white, fissured and rough.

Leaves: Opposite, 50-60 cm long, imparipinnate at apex and bi-or-tri-pinnate below; leaflets 4-6 cm long and 2-3 cm broad, ovate or ovate-lanceolate.

Flowers: Inflorescence 25-30 cm long, erect, terminal panicles; corolla tube 6-8 cm long, slender, cylindrical, white with greenish tinge, mouth nearly 2.0 cm across; stamens 4. It is a quick-growing tree and comes to flower in 4-5 years. Flowering occurs twice, in autumn and in spring, flowers open during night and fall down in early part of day. Some hang at end of thread-like pistils.

Fruits: A pod 30-45 cm long, flattened; seeds flat, thin, 2.5 cm long, with a broad, transparent delicate wing.

Propagation: It prefers a deep fertile soil and warm humid climate, but grows well in arid regions also. Besides seeds, tree is propagated easily from suckers. Tree may be multiplied from cuttings also.

Landscaping uses: Tree is extensively planted along road, in gardens and parks specially for its beautiful, fragrant flower. However, root system is comparatively shallow and wood is brittle.

Scientific name : *Mimusops elengi*

Common name : Elengi, Spanish cherry

Family : *Sapotaceae*

Elengi is the vernacular name of this plant in Malabar. Native to western Peninsula, the Andamans, Myanmar and Sri Lanka.

A medium-sized, handsome, evergreen tree with regular globular crown, reaching 10-15 m height; trunk round, straight, upto 3 m with numerous spreading branches; bark rather rough, dark grey. This tree grows in the tropical climate both in moist and dry regions with equal ease but cannot stand waterlogging.

Leaves: Alternate, upper side shiny green, lower side pale green, elliptic, shortly acuminate, 10-15 cm long, half as broad

Flowers: Borne in axillary cluster of 1-6; corolla gamopetalous, star-shaped, fragrant, off-white, 1.5 cm across; ovary silky pubescent, berry ovoid, 2-3 cm long, orange when ripe, containing 1 - 2 brown, polished seeds in a semi-sweet pulp. Flowering starts in spring and continues sporadically throughout the warm season.

Propagation: Easily propagated from seeds, the seedlings are rather slow-growing, coming to flower in 5-6 years. Tree assumes a perfect globular shape if grown without competition.

Landscaping uses: A very handsome tree for roadside planting and provides shade in gardens and parks. Occasionally grown in the vicinity of houses for its fragrant flower.

Scientific name : *Nyctanthes arbor-tristis*

Common name : Coral jasmine, Night jasmine, Tree of sadness, Tree of sorrow

Family : *Oleaceae*

Arbor-tristis is Latin means 'sad tree', coral jasmine is cultivated throughout the tropics for its lovely, scented flowers.

A large shrub or small tree reaching 5-7 m in height, often water-suckers come out profusely from the base, bark thick, rough and pale brownish and young branches quadrangular.

Leaves: Opposite, ovate with smooth or somewhat dentate margin, upto 10 cm long.

Flowers: Scented, in small dense clusters at the end of short stalk; corolla tube orange or orange-red, 1 cm; corolla lobe 5-8, of various shapes, 2.5-3 cm across, white. The tree flowers profusely in autumn, which open at night and fall off before sunrise; in milder climate stray flowers are observed in other seasons also.

Fruit: A flat, papery capsule, spreading into 2 flat parts, each with one seed.

Propagation: Easily propagated by seeds; the quick-growing seedlings may flower with good care in two years.

Landscaping uses: It is a very popular flowering plant in warn, humid regions and is planted in home gardens, religious places for the numerous scented flowers which fall off and form a carpet below in the morning.

Scientific name : *Parkia biglandulosa*

Common name : Parkia

Family : *Fabaceae*

Biglandulosa refers to two small glands at the base of the petiole or leaf stalk. Indigenous to the tropical forests of Assam (India), Myanmar, Malyasia. A large, upright, stately, showy, quick-growing, evergreen tree attaining a height of 20 m or more. The tree thrives in warm and moist climate.

Leaves: Pretty, feathery, bipinnate, 30 cm or more long, rachis of leaves downy; pinnae 45-50, each 8-10 cm long having 70-75 pairs of small, oblong leaflets about 1 cm long; inflorescence axillary, pendant, globular, 2-3 cm across, suspended by a long peduncle.

Flowers: Very small and insignificant, borne in heads which are at first brown and velvety, but white as the flowers open; upper flowers bisexual, lower ones male or sterile; stamens 10; filaments united at the base and adhering to the calyx tube; pods flat, coriaceous, dehiscent, containing some white, powdery mass. Flowering takes place in November-December followed by large, hanging clusters of long, brown pods in February-March.

Landscaping uses: The tree is valued for the feathery foliage and suspended inflorescence and pods. Suitable for planting in large gardens.

Scientific name : *Peltophorum inerme* syn. *P. ferrugineum*
Common name : Copper pod, Rusty shield-bearer, Braziletto wood
Family : *Fabaceae*

'Ferrugineum' in Latin means 'rust-coloured' in allusion to rusty colour of pod and branches. The species is a native of Sri Lanka, Andamans, Malaysia and North Australia. It is grown extensively in India and many other countries for its ornamental value. A partially deciduous big tree with grey bark, handsome foliage, reaching upto 20 m in height.

Leaves: Fern-like, bipinnate, 30-50 cm long, 20-30 cm broad, pinnae 4-13 opposite pairs, each one 5-15 cm long with 8-18 pairs, opposite, leaflets dark shiny green above, whitish below, nearly 2 cm long, 6-7 mm broad.

Flowers: 15-25 cm long terminal panicles, about 2.5 cm across; sepals 5; petals 5, roundish, crinkled, yellow; stamens 10, free, unequal. It is quick growing and flowers in 3 years under favourable conditions. In January, tree becomes nearly leafless. New leaves appear in spring along with flower. It flowers twice, profusely in March to June and a few in September to November.

Fruits: Flat, elongated, shield shaped, copper-coloured pod turning almost black on maturity, 4-6 cm long, 2-3 cm broad, indehiscent, woody, broadly winged containing 1-3 elongated seeds. Fruit remains on tree till next flowering.

Propagation: Tree grows equally well both in arid and moist localities upto 600 m elevation. Easily propagated from seed, cutting is also successful.

Landscape uses: A good shade tree and is largely planted along wide roads and in large gardens and parks. When alternated with *Delonix regia* which nearly has same blooming period, a pleasant contrast of red alternated with yellow brightens up landscape.

Scientific name : *Plumeria alba Linn.*
Common name : Pagoda tree
Family : Apocynaceae

This tree is distributed all over the country. It is a native of Mexico. It is smaller than temple-tree. It bears clusters of dark green ovate leaves, hairy below.

Flowers: Flowers have an exquisite fragrance and are white with no yellow colour in the centre. Blooms in March-April and again from July to October. It is the prettiest tree of *Plumeria* genus and is almost evergreen.

Propagation: Easily propagated from cuttings.

Landscape uses: An excellent tree for planting in compounds of houses.

Scientific name : *Plumeria acutifolia*
Common name : Pagoda tree
Family : Apocynaceae

The *Plumeria* is named after a French botanist, Charles Plumier; 'acutifolia' refers to sharp pointed or tapering leaves. A native of Mexico and Guatemala, it is grown extensively all over the country. It is a small deciduous tree with a crooked stem. Bark produces a milky juice when broken.

Leaves: Leaves are big, broadly lanceolate, tapering at both ends and quite distinctive due to prominent parallel veins. They are borne in crowded spiral at the end of long branches. Tree remains leafless from

December until rains. Young trees, however, remain leafy and attractive throughout the year.

Flowers: Flowers are waxy with white yellow centres (*Plumeria acutifolia* var. *alba*), and red in colour (*Plumeria acutifolia* var.*rubra*) highly fragrant and borne in upright clusters. Petals are sometimes tinged with pink below. Flowering period is March-April, and again from July to October. Tree claims great affection because of its sweet scented flowers which bloom practically throughout year.

Fruits: Appear as two horn like pods on a long pendent stalk and seeds with a tassel of silk. Tree rarely fruits or seeds in India. It is a highly ornamental plant and is often associated with temples.

Propagation: Easily propagated by cuttings in July and August. Cuttings should be allowed to wilt before planting. It thrives in sandy soil with a rainfall of 80 cm upwards. In India, tree is venerated as the "Tree of Life". Even if its roots are exposed, it will still go on growing and blooming, without showing any sign of wilting. If replanted in a new location, tree again grows as if it had never been subjected to any harsh treatment.

Scientific name : *Polyalthia longifolia*

Common name : Mast tree

Family : *Annonaceae*

Longifolia in Latin means 'long leaf'. A native of Sri Lanka and adjoining areas of South India. It is grown extensively as ornamental plant throughout the tropics. The tree can grow on any soil and climate except on deserts, upto 100 m elevation. It can stand drought and hot climate very well and also grow vigorously in humid regions. Crown up plant stands waterlogging for quite some time.

A tall, deciduous, handsome tree of pyramidal shape with dark bark, a tall straight stem and slender branches.

Leaves: Lance-shaped, glossy green, translucent, narrowing to a point, with wavy or undulated margin, nearly 15-25 cm long and 3-4 cm wide, placed alternately on the joining branches. Tree becomes leafless towards the end of winter; new leaves arise shortly which are glossy light green in colour.

Flowers: Greenish yellow, star-shaped, bi-sexual, 3 cm across, on long slender pedicels, in sessile or shortly pedunculate umbels, borne on one-year old twigs in profusion; sepals 3, greenish, broad, short; petals 6, narrowly lanceolate, spreading. Flowers which appear in abundance along with leaves in March-April are not conspicuous, because the colour of new leaves and flowers cannot be distinguished from a distance. Fruits ripen in July-August.

Fruits: Cluster of many pedunculate carpels on a common stalk, egg-shaped, about 1 cm long, rounded at both ends; seeds smooth.

Propagation: Propagated by sowing fresh seeds.

Landscaping uses: These trees are favourite as avenue plants. They are also suitable for line planting and screening.

Scientific name : *Pongamia pinnata* syn. *P. glabra*
Common name : Indian beech, Ponga oil plant
Family : *Fabaceae*

Pinnata in Latin meaning 'feathered'. Common around the margin of inland water bodies and also in the tidal and beach forest of India, Sri Lanka, Malayan Archipelago, extending to the coasts of South China, the Fiji islands and tropical Australia.

A moderate sized, spreading, deciduous tree reaching 10-12 m with smooth, greyish-brown bark, a short trunk and rounded crown. Although *Pongamia* grows in all types of soil in tropical region, it prefers moist soil and warm humid climate.

Leaves: About 25 cm long, pinnate; leaflets shiny dark green, ovate or elliptic, 10-12 cm long and 6-8 cm broad. Tree becomes leafless in the early spring and flowering occurs along with the emergence of new leaves.

Flowers: Rose or lilac, about 1.2 cm long in short racemes in the axils of new twigs. Flowering starts at the age of 3-4 years. The ground below the tree becomes thickly covered with fallen flowers.

Fruits: Pods woody, non-dehiscent, rhomboid, 4-5 cm long, 2-3 cm broad containing usually 1-3 seeds.

Propagation: Fresh seeds germinate readily.

Landscaping uses: It is suitable for planting along the roads and highways, banks of canals and pools and in parks and large gardens. It has medicinal poperties.

Scientific name : *Pterospermum acerifolium*

Common name : Muchkunda, Maple twist, Kanak champal

Family : *Sterculiaceae*

Tall, evergreen tree with smooth, grey bark, twigs covered with rusty brown hairs, often reaches a height of 12-15 m

Leaves: Tough, thick, leathery, variable in shape and size, 12-20 cm long and 10-15 cm broad, orbicular, shallowly and irregularly lobed, glabrous above, silvery below.

Flowers: Hermaphrodite, large, solitary or in pairs in axil of leaves, highly fragrant; calyx composed of 5 linear, thick, fleshy, segments upto 12 cm long, curling backwards as flower opens; petals 5, linear-oblong, pure white; staminal column short, divided into 20 filaments; ovary 5-celled. Flowering occurs in spring and early summer and flowers retain scent long after they are dry.

Fruits: woody capsule, 10-15 cm long, 5-8 cm wide, oblong; seeds numerous, oblique with brown, papery wing. Fruits take about a year to mature, when they split releasing winged seeds to float away in air current.

Propagation: Easily propagated from seeds, tree is a quick grower, producing flowers from third year under favourable conditions. It prefers warm climate and high rainfall.

Landscape uses: This is a good tree on medium roads and in large gardens and parks.

Scientific name : *Saraca asoca* syn. *S. indica*

Common name : Asoka tree

Family : *Fabaceae*

The species got its name from Sanskrit *asoka*. Tree is found wild in India and also native to Sri Lanka, Myanmar and Malyasia. It is widely grown in gardens, particularly adjoining temples and monastery for its ornamental value and flowers. A handsome, medium sized, spreading, evergreen tree with rounded crown, reaching 6-8 m; trunk erect, short, covered with a smooth, dark brown bark; young shoots and leaves drooping and coloured. *Saraca asoca* is hardy and grows well in warm, humid climate of tropical and subtropical regions. It needs shelter from dry wind, prefers a semi-shady situation and a porous, moist soil.

Leaves: Alternate, paripinnate, 25-30 cm long; leaflets opposite, 3-6 pairs, elliptic oblong, 10-20 cm long and 5-7 cm broad, glabrous; young leaves faccid, greyish white quickly changing to purplish red, finally green and stiff.

Flowers: Flowers in dense, orange-red, corymbose cluster, 10-15 cm across; calyx tube about 1 cm long; opening out into 4 to 6 roundish petaloid lobes, at first orange, turning to scarlet afterwards; 1.5 cm across, petals absent; stamens perfect, filaments 6-8, hairy, 2.5-3 cm long, at first orange near base and scarlet upwards; Tree comes to flower when 4-6 years old. Major flowering season is from March to May. Tree looks marvellous when bright scarlet, ixora-like blooms are produced profusely.

Fruits: Pod dehiscent, 10-20 cm long, 5-7 cm wide; Seeds 4-8.

Propagation: Easily propagated from seed.

Landscaping uses: This tree is very suitable for all gardens and parks, big and small, young foliage very attractive, used as an accent tree.

Scientific name : *Spathodea companulata*

Common name : Scarlet bell tree, African tulip tree, Syringe tree, Fountain tree

Family : *Bignoniaceae*

'Companuata' in Latin means 'bell-shaped', for appearance of the corolla. Widely grown as a flowering tree in tropical region of Asia and America. A handsome, evergreen or deciduous tree in dry area, reaching 15-20 m, with light grey, fibrous bark and short branches; young shoots hairy. Tree grows in a rich, light soil upto 1000 m elevation in tropical and subtropical climate.

Leaves: Densely arranged near tip of branches, opposite, odd-pinnate, 30-45 cm long; leaflets 9-17, ovate-lanceolate or oval, pointed, 8-10 cm long and 5-6 cm wide with short stalk.

Flowers: Inflorescence large, many-flowered terminal clusters; sickle-shaped buds filled with liquid; calyx boat shaped, 4 cm long; corolla bell shaped with 5-8 triangular lobes, orange-crimson outside, about 10 cm long having yellow wrinkled margin. Flowering usually starts in January-February. In moderate climate tree continues to flower for many months, major flushes appear in summer.

Fruits: Long, pointed pod, 20-25 cm long, 4-5 cm broad, containing many thin papery seeds.

Propagation: It is easily propagated from seeds. Root suckers and cuttings are also successful. Seedlings grow quickly and begin to flower when 3-4 years old. In dry regions tree is deciduous.

Landscaping uses: Suitable for medium and large gardens and avenues.

Scientific name : *Tabebuia rosea* syn. *Tecoma rosea*
Common name : Rosy trumpet tree
Family : *Bignoniaceae*

Rosea is allusion to rosy colour of flowers. A native of Mexico, this ornamental tree is planted widely in gardens of tropical world including India. A tall, erect, deciduous tree reaching a height upto 20 m. It prefers warm and dry climate and grows on most of soils.

Leaves: Opposite, palmately compound; leaflets 3-5, stalked, 7-12 cm long, 3-5 cm wide. In most cases, leaves fall off in winter and numerous rosy-pink flowers appear on leafless trees; Leaves arise after main flush of flowers.

Flowers: Large, shades of pink, 5-lobed, wrinkled; stamens 4. Flowering duration is a fortnight. In moderate climate, flowering continues for several months.

Fruits: A linear, dehiscent capsule, 12-15 cm long; seeds numerous, flat, winged.

Propagation: Propagated from seeds, seedlings grow fairly fast and comes to bear in 4-5 years.

Landscaping uses: This is a good roadside tree and suitable for growing in parks and other public places.

Scientific name : *Tabebuia spectabilis*

Common name : Tabebuia

Family : *Bignoniaceae*

A small to medium tree, of 15 m high. Native of Colombia, grown as ornamental plant in tropical and subtropical countries.

Leaves: Digitate with 5-6 leaflets; leaflets ovate to ovate-oblong, 6-12 cm long.

Flowers: Inflorescence terminal panicles of bright yellow, funnel shaped, 4.5-5.5 cm long, corolla lobes fringed, 4 cm across. Flowers are produced in March-April.

Scientific name : *Thespesia populnea* syn. *Hibiscus populneus*

Common name : Bhendi tree, Portia tree

Family : *Malvaceae*

'Populnea' refers to its poplar-like leaves. Tree grows naturally in beach and tidal forests of India, Bangladesh, Myanmar and other tropical countries. A small, evergreen tree with umberlla-like, dense, hemispherical crown, reaching 8-10 m; bark brown, rough and often

knobby. It requires a moist tropical climate and porous soil, can stand salinity well and grows luxuriantly along sea coast. However, it is grown throughout India except arid regions where growth is not satisfactory.

Leaves: Green, cordate, acuminate, entire, 8-12 cm long, 6-10 cm wide; veins whitish, prominent against green background.

Flowers: Axillary, mostly solitary, large, 5-8 cm across, turning purplish with age, dark maroon centre in one type, self-coloured in other, resembling tulip; calyx truncate, bell-shaped, shallow.

Fruits: A capsule, oblong, flattened at the top, 2-3 cm across, 5-valved, containing many silky seeds.

Propagation: Easily propagated from seeds or from cuttings, large ones being preferred. It is a quick-growing tree coming to bloom in 2-3 years. In place away from sea coast, it is rather a short-lived tree which usually dies in 30-40 years. Tree is in bloom more or less throughout year, mass being produced during April-May and July-September.

Landscaping uses: It is a good shade as well as flowering ornamental tree for planting along road and in private gardens and parks.

Economic trees

Scientific name : *Ailanthus altissima*

Common name : Tree of heaven

Family : *Simaroubaceae*

A big, deciduous tree with an erect trunk, almost smooth grayish bark and white, compact but rather light wood; the wood and bark contain tannin and other astringents. The branches are spreading and young branchlets are covered with a yellowish tomentum. The alternate leaves, unevenly pinnate, are composed of 13-15 lanceolate-acuminate leaflets; they have no stipules. Oil-bearing glands at the base of the leaf give off a disagreeable smell if rubbed, as do the flowers. The plant is polygamous, with small, yellowish-green, five-petalled flowers, male or hermaphrodite; the bisexual ones have a central pistil with 5-6 carpels and 10 stamens, of which 5 are relatively short. The inflorescences are long, dense panicles. The fruit is a reddish, winged samara.

Propagation: By seed or by basal suckers, which grow easily; the tree seeds freely disperse and can become a pest.

Scientific name : *Anacardium occidentale*

Common name : Cashew nut

Family : *Anacardiaceae*

An erect, spreading, evergreen tree, growing up to a height of 15 m, with rough bark. Leaves glabrous, thickly coriaceous, obovate, 10-16 cm long, entire. Flowers polygamous, 1-1.5 cm across, in large terminal bracteate panicles, pentamerous; petals linear, pale yellow, streaked with pink; stamens 10, one longer and always fertile, the rest of ten sterile. Fruits a drupe, kidney-shaped, 2-3 cm long, seated on fleshy, thick pseudocarp, 5-8 cm long, orange or crimson.

This tree is common in tropical America, from Mexico to Peru and Brazil. It was introduced into India from Brazil by the Portugese some 400 years ago and has become naturalized in the West Coast. It is also cultivated in the coastal regions of South Africa, Malagasy, Mozambique, the West Indies, and is found in South-East Asia from Sri Lanka to the Philippines.

The tree is usually grown from seed and it can also be propagated by grafting and layering. The tree begins to flower in December and flowering continues for about 3 months. It generally takes 2-3 months before the fruits are mature enough to be gathered, and only the mature ones are collected before they fully ripen and fall to the ground. The season for collection extends from the end of February to the end of May or early June.

Fleshy peduncles of the fruit is known as 'cashew apple', while the fruit is the kidney-shaped nut attached to it. The white kernels in the nuts possess pleasant taste and flavour. They are eaten either raw or fried and salted or sugared, also used in large quantities in the preparation of sweetmeat and in confectioneries. They constitute a highly nutritious and concentrated food. The commercial shell oil obtained by the roasting of cashew nut is a dark brown viscous oil, easily soluble in most of the organic solvents, and the oil is used as a preservative in the painting of boats, fishing nets and many other industrial purposes. Cashew apple is edible and yields a delicious beverage. The juice is fermented and made into a wine which retains the flavour of the fresh fruit. The tree yields a pale yellow to reddish gum.

Scientific name : *Bixa orellana*

Common name : Annatto tree, Sinduria

Family : *Bixaceae*

A little tree with alternate, simple leaves, with stipulcs and a long stalk. The inflorescence, formed by hermaphrodite flowers with a deciduous calyx, is a long, low panicle of a beautiful flesh pink. The fruit, which is what the tree is planted for, is a capsule with soft red skin and red juice. The dye obtained from it, known in the trade as annatto or 'arnotto', is used to dye fabrics orange-red. An oily solution of the yellow colouring agent contained in 'annatto' is the only dye permitted by law for colouring butter and cheese.

Scientific name : *Camellia sinensis*

Common name : Tea tree

Family : *Theaceae*

These are small trees. In the wild state they will grow as much as 12-15 m, but on plantations they are kept down by continuous pruning to a height of 1-1.5 m. They have leathery elliptical lanceolate leaves with finely toothed edges, shiny dark green, with a short stalk. The little flowers, white and scented with numerous yellow stamens, are solitary or in cymes of 2-5; the fruits are capsules containing 1-4 seeds. The plant is pruned for two purposes; to make picking easier and to encourage the production of young leaves, which are picked every three weeks from the third year onwards. There is a difference between green teas, prepared by simply cutting the leaves, and black teas, which are obtained by fermenting the leaves.

Scientific name : *Ceiba pentandra*

Common name : Kapok tree

Family : *Bombacaceae*

A tall tree, as much as 30-40 m high, with a stout, spiny trunk with big buttress formations at the foot; the branches, thick and strong, are wide spreading and almost horizontal, arranged in layers. The compound leaves are digitate, with 5-7 entire leaflets. The flowers are white and large, growing in little clusters, with a gamosepalous calyx and oblong petals from which the staminal column protrudes. The fruit is an oblong, leathery capsule, about 10 cm long, which opens in five sections to release a great number of brown, ovoid seeds contained in masses of a thick fiber rather like cotton, this is the kapok that is used for stuffing pillows.

Scientific name : *Cinchona calisaya*

Common name : Quinine tree

Family : *Rubiaceae*

An evergreen tree up to 30 cm high with erect trunk and yellowish bark. The leaves are opposite, oblong and leathery, with reddish stalk and veins. The flowers, pink and scented, grow in panicles and the fruits are small ovoid capsules. The bark of the tree is rich in alkaloids, the most important being quinine, which is effective in the treatment of malaria.

Scientific name : *Cinnamomum camphora*

Common name : Camphor tree

Family : *Lauraceae*

An evergreen tree that reaches a height of 15 m. Its leaves are pale green, glaucous underneath, alternate, ovate-acuminate and leathery, with three distinct veins; in the juvenile stage they are rosy. The little whitish-yellow flowers grow in axillary panicles, and the fruits that follow them are small ovoid drupes, dark purplish-red when ripe. Camphor occurs in the essential oil contained in all the tissues of the tree. If the leaves are rubbed they smell of camphor-but mostly in the peal of wood, both that of the aerial parts and of the roots, where it accumulates particularly in hollows. It is used to be extracted by felling the trees, but nowadays the leaves and young branches are also used and the trees are not cut down until they are 50 years old. The product, widely used in the manufacture of celluloid and as a preventive against parasites is useful in medicine. Nowadays it is also obtained synthetically, since the natural substance has become so costly.

Scientific name : *Cinnamomum zeylanicum*

Common Name : Cinnamon

Family : *Lauraceae*

An evergreen tree up to 6-10 m tall, with erect trunk and pale brown bark; young branches are almost rectangular in section. The leaves about 15 cm long, are leathery and stiff, opposite, lanceolate-ovate, bright green on the upper side, pale or glaucascent below, and furrowed by three strong veins. The flowers are small and yellowish white, with a tubular corolla and six wide lobes, velvety at the edges, they are clustered in axillary and terminal panicles. The fruit is a berry with a single seed, which remains within the persistent calyx and turns black when ripe. To obtain the spice in plantations, the bark is removed and dried. A yellowish-brown oil is prepared from it which is used as a flavour and in medicine.

Scientific name : *Citrus aurantium*

Common name : Bitter orange /Seville orange

Family : *Rutaceae*

A tree that may grow to a height of 15 m with a diameter of 50 cm. It differs from the Sweet orange in the long thorn at the axil of the lower leaves, the darker, more aromatic foliage, the more broadly winged stalk, the rougher and more brightly coloured peel on the fruit, in the sour, bitter flesh. The peel is used to make orange bitters, used in cocktails, and sweet liquors such as Curacao. The strongly scented flowers are gathered by shaking the trees so that they fall into cloths placed below the branches. Fruits are distilled to make an essential oil known as neroli oil; very valuable in perfumery. Another essential oil, 'petitgrain', is obtained by distilling the leaves and twigs.

Scientific name : *Citrus bergamia*
Common name : Bergamot
Family : *Rutaceae*

This is a small tree, growing to 3 m with pendulous branches and rather pimply leaves with slightly winged stalks. The flowers with five green sepals, five white petals and ten stamens are scented. The fruit, a hesperidium, is spherical, a little larger than the oranges, with a thin lemon-yellow skin; it contains 10-15 segments and has sour, inedible flesh. It is grown exclusively for the essential oil pressed from the skin. This essence is used in the manufacture of liquors and perfumes, including Eau de Cologne.

Scientific name : *Coffea arabica*
Common name : Coffee tree
Family : *Rubiaceae*

A small evergreen tree 2-3 m in height, with spreading, flexible branches; the opposite leaves are shiny green, oblong-acuminate, with a short stalk and very marked veins. The white, scented flowers with a tubular corolla, appear in clusters at the leaf axils and are followed by the fruit, drupes with two seeds and a fleshy covering, green at first, red when ripe. Each of the two seeds is convex on the outside and flat, crossed by a furrow on the inside; these are the so-called coffee beans which are roasted.

Scientific name : *Cola acuminata*
Family : *Sterculiaceae*

This tree is about 20 m tall, has persistent stalked leaves, alternate, with a leathery lanceolate-acuminate blade with prominent veins on the underside; in the juvenile stage they may be deeply incised toward the base. The flowers, unisexual or polygamous, grow in dense axillary or terminal panicles and have no corolla at all; the five yellow, showy segments, with purple stripes at the base, are only the opened lobes of the thin, tubular, greenish calyx. The fruits, about 15 cm long, are green, oblong and leathery, made of 4-6 sections which are the so-called cola nuts, appearing while unripe as tuburcles within the outer covering. They are formed by the cotyledons and are irregular in shape; when fresh they vary from white to yellow or pink, when dried they are reddish. They contain some 2% caffeine and a little theobromine and other drugs. The local people use them as stimulants, and large quantities are exported as ingredients of carbonated soft drinks.

Scientific name : *Dalbergia sissoo*

Common name : Sissoo

Family : *Fabaceae*

This is a big tree which in its native surroundings can grow to more than 20 m. The leaves are imparipinnate, with five leaflets, alternate and stalked, sharply acuminate and downy on the underside. The flowers are white and papilionate, growing in small panicles; the fruit is a pod. This tree, besides being grown as an ornamental for its large crown, which often starts comparatively low, has been introduced into all tropical and subtropical regions on account of its commercial value. Its timber elastic and easy to work, being used for all kinds of work calling for strength and durability, including ship building.

Scientific name : *Dracaena draco*

Common name : Dragon tree

Family : *Agavaceae*

This slow-growing, strangely shaped tree belongs to the monocotyledons. Though, having a woody stem, it is often included in the Liliaceae. The thick trunk, which becomes branching, has apical clusters of sword like leaves arranged in rosettes, and can grow 20 m high and over 4 m in diameter. It is an exceptionally long-lived tree; two specimens in the island of Tenerife are believed to be among the oldest trees in the world. The flowers growing in big panicles, are bell-shaped, yellowish-white; they appear at the ends of the branches inside the leaf rosettes and are followed by orange berries. The resinous reddish liquid that oozes from the trunk is one of several resins called dragon's blood which used to be of great commercial importance as the base of fine dyes and paints and in photo engraving until the invention of synthetic colours.

Scientific name : *Erythroxylon coca*

Common name : Coca

Family : *Erythroxylaceae*

A small tree that reaches a height of 3 m and begins to branch in the juvenile stage, though it has rather slender trunk and branches. The little, short-stalked elliptical leaves are alternate and entire, brilliant light green and more or less acuminate according to the variety. The small flowers have a green calyx and five white petals; they grow on stalks at the leaf axils, solitary or in twos or threes, and are followed by the fruits, small drupes with a brilliant red aril. The alkaloid from which cocaine is prepared is extracted from the leaves on an industrial scale.

Scientific name : *Euclayptus globulus*

Common name : Blue gum

Family : *Myrtaceae*

An evergreen tree that can reach 40 m and more in height, with deciduous bark which comes off in long strips, leaving the trunk patterned in silver-gray. The young leaves are round, sessile and opposite, but on maturity they become lanceolate, stalked, alternate and aromatic. The flowers are solitary, sometimes growing two or three together, with short stalks; they consist of a calyx in the form of a truncated pyramid and an operculum formed from four petals joined together, which falls at maturity. When the operculum falls, the many stamens and the style emerge to give the flowers a feathery look. Flowering occurs between autumn and spring and is followed by the ripening of the fruits, which are angular capsules with a great number of very light seeds. The wood is used mainly for cellulose.

Scientific name : *Haematoxylon campechianum*
Common name : Compeachy tree/Logwood tree
Family : *Fabaceae*

A tree 6-9 m high with spiny branches and compound, pinnate leaves consisting of 2-4 pairs of wedge-shaped leaflets deeply incised at the edges. The flowers are small, yellowish and scented. The fruits are oblong pods. The wood has a heavy, hard heartwood, red-brown to violet, from which an active principle called haematoxylin is extracted - a colourless substance which is transformed on oxidation into a red dye, haematin, which can react with metallic hydrates to produce coloured lacquers. Fragments of the wood treated with steam under pressure give campeachy extract, a mixture of haematoxylin and haematin used for varnishes and inks and to make a dye for silk, wood and cotton. Haematoxylin is used as a colouring agent for microscopy.

Scientific name : *Hevea brasiliensis*
Common name : Hevea rubber
Family : *Euphorbiaceae*

A tropical tree that reaches a height of 18 m and is the principal source of rubber, which is obtained by treatment of its coagulated latex. The leaves are alternate, with long stalks, composed of three lanceolate-acuminate leaflets. The plants are monoecious, with small flowers, pubescent and lacking petals, clustered in panicles. The fruit is a capsule. The latex is collected by making an incision in the bark, which is light gray.

Scientific name : *Ilex paraguariensis*
Common name : Mate
Family : *Aquifoliaceae*

A small evergreen tree that reaches a height of 6 m. The leaves are leathery, lanceolate or oblong with a narrow base, a flattened tip and rounded teeth at the edges. The small, white flowers grow in clusters at the leaf axils. The fruits are red, globoid berries with four seeds. The plant is grown for its leaves, from which a drink is made that is famous all over South America; it is sometimes called "Jesuits' tea" because when they returned from their missions the Jesuits were the first to make the drink known in Europe. The dried leaves are roasted and pulverized; hot water is poured on the dry powder to make a drink that is stimulating. The local people drink it through straws perforated at the tip, after preparing it in specially shaped wooden gourds, known as mate gourds.

Scientific name : *Liquidambar styraciflua*
Common name : Sweet gum
Family : *Hamamelidaceae*

A deciduous tree which in its native conditions reaches a height of 40 m, with a tall, narrow crown; the bark is deeply fissured even on young branches. The alternative leaves are stalked and palmate, with 5-7 acuminate lobes, smooth and glossy on the upper side, downy along the veins on the underside. Their bright green leaves change to vivid scarlet before they fall. The flowers are unisexual, growing on the same plant, insignificant and apetalous, borne on round inflorescences. The fruits are really round, made up of a number of dehiscent capsules each with 1-2 winged seeds and prominent spines formed from the persistent styles. Both wood and leaves are aromatic. When the bark is stripped, boiled and pressed it yields a resinous oil called 'storax' which is used as a fixative in perfumes and in therapeutic preparations.

Scientific name : *Moringa oleifera*
Common name : Horseradish tree
Family : *Moringaceae*

A small tree that grows up to some 8 m with a trunk of softwood and corky bark; young shoots are pubescent. The leaves are tripinnate, with no stipules, as much as 50 cm long, with stalked leaflets. The flowers, growing in panicles, have a gamosepalous calyx with five deflexed lobes and a corolla with five petals, four deflexed and one erect with the anthers joined to it, while the stamens have long, straight filaments. Individual flowers are whitish, pedunculate and perfumed, followed by fruits that are capsules up to 30 cm long, with nine big triangular seeds with three wings. An oil is extracted from these which never goes rancid. It is used for delicate precision machinery and is known as 'watchmaker's oil', or sometimes 'behen oil'. The roots, too, which have a piquant taste, are sometimes used as spices and condiments.

Scientific name : *Myristica fragrans*

Common name : Nutmeg

Family : *Myristicaceae*

An evergreen tree 10-20 m high, with alternate leaves, oval with a short stalk, dark green on the upper side, lighter on the underside, and aromatic. The plants are dioecious, so that it is necessary to grow at least one tree with male flowers near several with female flowers. The inflorescences are axillary, the flowers, small, yellowish or greenish; the males have a number of stamens joined in a little column, the females have a single carpel, and both are scented. The fruit that follows fertilization is a fleshy, pear-shaped berry, which opens on ripening to release the big seed, surrounded by a fibrous aril which changes from reddish to yellowish when dried. Marketed under the name of mace, it is used in cooking and perfumery. The bare seed, generally treated with lime, is the nutmeg used to flavour food and drinks.

Scientific name : *Olea europaea*

Common name : Common olive

Family : *Oleaceae*

Growing to 8 m, it develops a large, twisted trunk covered with smooth gray bark which cracks into scales with age. The crown is about as wide as the tree is high. In good soils it is generally open and asymmetrical; in poor soils it is dense and rounded. The leaves are lanceolate or oblong, 2.5-7.5 cm long. They are leathery, gray-green above, silvery beneath. Small yellow-white, fragrant flowers are borne in short racemes on the previous season's growth. These are followed by the familiar fruits, which are actually fleshy, oily drupes. The olive tree is also an externally handsome ornamental.

Scientific name : *Pimenta dioica*

Common name : Allspice, Pimento

Family : *Myrtaceae*

A little tree that in the countries of origin may reach a height of 10 m but which remains lower when grown under glass. It has a flexible trunk with silver-gray bark and many thin branches, with leathery, opposite leaves, oblong-acuminate, with strong veins. The little, whitish flowers grow in cymes, generally apical, sometimes axillary and have 4-5 petals; the calyx has four lobes, which distinguishes this from other species. The globose fruits are very small, only about 3-4 mm. It is for this spice that the plant is cultivated. It is used as a flavouring for foodstuffs and the scent of it seems to combine several aromas, so that it has acquired the name 'allspice'.

Scientific name : *Quercus suber*

Common name : Cork oak

Family : *Fagaceae*

This tree, with its twisted trunk, does not exceed 15-20 m in height. It has a thick, deeply cracked bark, whitish outside and red inside. The irregular branches form a broad, not very symmetrical crown, of leathery evergreen leaves, lanceolate with slightly toothed margins, of a beautiful dark glossy green on the upper side and white and downy below from the presence of stellate hairs. The flowers, which appear in April-May, are of separate sexes; the males in long, pendulous catkins, the females in erect spikes with 2-5 flowers. The fruit is a typical acorn, the cup made of gray, downy scales. The basic importance of this plant is due to the exploitation of its thick, corky bark, which is stripped every 9-10 years as required, leaving the distinctively rusty-coloured trunk bare.

Scientific name : *Rhus verniciflua*
Common Name : Varnish tree
Family : *Anacardiaceae*

A tree up to 25 m high, the tallest of its genus, with big, imparipinnate leaves carried on a short stalk and formed from 11-15 ovate-oblong or oblong-lanceolate leaflets 12 cm in length, entire, rounded or pointed at the base and downy on the underside. The fruit is a drupe thicker than it is long, and yellowish. The sap is a poisonous, sticky juice which oxidizes on contact with the air. For use as a vitrifying lacquer it must be applied with the brush before it hardens, which for no known reason happens better in humid conditions. When set the lacquer becomes extremely hard, so that it can be polished to a high level of brilliance. The extraction of this sap has been practiced in China since prehistoric times.

Scientific name : *Simarouba glauca*
Family : *Simaroubaceae*

A medium-sized tree, up to 15 m high; bark grey, smooth. Leaves 20-40 cm long; leaflets 11-21, oblong or oblong-elliptic, 5-10 cm long, rounded, retuse or abruptly pointed at the apex, cuneate at base, glabrous, coriaceous, dark green above, pale or glaucous beneath. Panicles 10-30 cm long, widely branched, the staminate more copiously flowered than the pistillate; calyx 3-3.5 mm wide; lobes ovate to triangular; petals oblong to obovate, 4-6 mm long, rounded, pointed at apex. Drupes ovoid or oblong, 1.5-2 cm long, slightly oblique, scarlet or dark purple.

Indigenous in Cuba and distributed in the West Indies, Florida, southern Mexico to Panama, northern South America, introduced and widely grown in many tropical countries. The tree is fairly fast growing and it is not browsed by cattle. It flowers in March-April.

It is grown for fuel wood; oil from kernels and for establishing vegetation in degraded land exposed to long spells of drought. The fruits are edible.

Scientific name : *Swietenia mahagoni*
Common name : Mahogany
Family : *Meliaceae*

An evergreen tree which grows more than 30 m high, with a graceful trunk and pinnate composed of 4-5 ovate-acuminate leaflets. The flowers are small and yellow-white, growing in axillary panicles, and are followed by large fruits which open at the base to release a number of winged seeds. The wood, for which the tree is highly valued, is in very great demand both for its beauty and for its hardness; it has been used since 1500 for quality carpenter's work. The sapwood is yellow, while the heartwood has a beautiful brown colour. If they are well grown, the branches are often preferred to the trunk because of their close grain. The tree is now growing rare, and the name mahogany is given to less valuable woods with similar properties.

Scientific name : *Syzygium aromaticum*
Common name : Clove tree
Family : *Myrtaceae*

A tree 10-12 m high, with a grayish trunk and erect ramification, giving it a conical or cylindrical shape. The persistent leaves are ovate acuminate and leathery, with conspicuous stalk and veins. The flowers are arranged in terminal corymb like cymes, three to each branch of the inflorescence; they have a yellowish-red gamosepalous calyx, long and cylindrical, with five lobes, red petals and a number of projecting stamens. The flower buds are picked and dried while the petals are still closed, looking like a little light coloured ball on top of the calyx; these are the cloves used as spice and in medicine.

Scientific name : *Tamarindus indica*
Common name : Tamarind
Family : *Fabaceae*

A big tree as much as 25 m in height, with a trunk about 7 m in circumference, with gray-brown bark. The pinnate leaves have 10-20 pairs of leaflets; the flowers grow in big racemes and have three upper petals, yellow veined in red, and two small lower petals; the staminal tube is long and down-curved. The fruit is a long, curved, indehiscent pod, slightly flattened between the fleshy seeds, which are surrounded by a thin membrane. Their flesh is rich in sugars and acids, the source of ingredients for drinks and substances used in medicine; the bark, and even the flowers, are also used for pharmaceutical purposes. The wood is strong and easy to work.

Scientific name : *Tectona grandis*
Common name : Teak
Family : *Verbenaceae*

In its native forests this tree can reach a height of 60 m. It has a slender trunk, a round crown and ovate leaves up to 60 cm long and 30 cm across, very downy on the underside. The bluish-white flowers grow in panicles 35 cm across, but many of the flowers are sterile. The plum like fruits have few fertile seeds; they are heavy and compact, offering little for the wind to catch hold of. The grayish bark surrounds a light sapwood that is vulnerable to termites and rot caused by fungi; the heartwood, in contrast, is immune, and during seasoning changes colour from yellowish to chestnut with darker streaks, retaining some of its pleasant smell. The wood of this tree, known commercially as teak, is widely used not only for its durability and hardness but above all for its low coefficient of contraction; it does not absorb moisture and is easy to work and polish. It is used for making furniture, parquet floors and veneers.

Scientific name : *Terminalia catappa*
Common name : Myrobalan/ Indian almond/ Tropical almond
Family : *Combretaceae*

A deciduous tree 20-25 m high, with the branches growing horizontally on the main axis, in whorls. The bark is brown-gray. The leaves, about 30 cm long, are obovate and blunt at the tip, and grow at the ends of the young twigs; they have a raised venation, especially on the underside, where they are pubescent. The flowers, small and whitish, with yellow anthers, are clustered in recurved terminal spikes. The fruit is an oval, indehiscent drupe 5 cm long, with a fleshy outside and a woody interior, rather like an almond, which contains the seed. It is regarded as edible in India, and an oil is also extracted from it, but the main value of the fruit lies in its content of up to 50 per cent tannins, which makes it much in demand for tanning skins.

Scientific name : *Theobroma cacao*
Common name : Chocolate tree/ Cacao/ Cocoa
Family : *Sterculiaceae*

An evergreen tree, 6-8 m high, with pubescent twig. Leaves alternate, oblong-ovate or elliptic-oblong, 15-30 cm long, abruptly acuminate, rounded at base. Flowers small, in fascicles, directly on the bark of the trunk and main branches, about 2 cm across when expanded; calyx rose-coloured with acuminate lobes; corolla yellowish; petals long with a stalk-like claw and expanded limb. Fruits elliptic-ovoid, up to 25 cm long, about 10-ribbed, purple-red, yellowish purple or brown; cells 5 with 5-12 beans in a row embedded in white or pinkish acid pulp. Though a native of the West Indies and South Central America, it is grown as plantation crop in many tropical countries of the world.

Good growing conditions for cacao are associated with high humidity and it is influenced by the rainfall and temperature. In favourable climate, cacao flowers and produces fruits during the greater part of the year.

The seeds are a source of beverage. They are called cacao beans and are the source of cacao, chocolate, and cacao butter, widely used in confectionery, milk chocolates, powdered chocolate etc. Raw seeds contain vitamins of B group; shells of fermented sun-dried beans contain appreciable quantity of vitamin D.

Scientific name : *Thespesia populnea*
Common name : Tulip tree/ Portia tree/ Umbrella tree
Family : *Malvaceae*

A medium-sized, evergreen tree, to 16 m high, with well-shaped crown, if grown undisturbed; branchlets peltate scaly; bark rough and brown. Leaves cordate, 10-15 cm long, acuminate, entire, covered with minute peltate scale, petioles long. Flowers funnel-shaped, solitary, axillary, 5-7 cm across, yellow with a deep red centre fading to dull purple; calyx cup-shaped, truncate; stamens numerous, form a tube. Fruit a capsule, oblong, depressed at the centre above, smooth, 2.5-3.5 cm in diameter, 5-valved; seeds silky.

The tree is distributed in the beach and tidal forests in many tropical countries including India, Sri Lanka, Myanmar, Malaysia, the Paciifc Islands, Africa, often grown on roadside and in gardens.

The wood is hard and durable and used for making boats, carts etc. The tree is largely cultivated for shade and large flowers. The tree is very valuable for its medicinal uses. It is mostly used as a remedy for skin diseases and leaves, flowers, fruits and bark are also employed. The root is used as a tonic, bark cures dysentery, fresh fruits bruised and applied to foreheads to cure migraine.

Medicinal trees

Scientific name : *Abroma augusta*

Common name : Devil's cotton/ Perennial Indian hemp

Family : *Sterculiaceae*

This is a small tree or shrub with velvety branches. Leaves are repand-denticulate.

Flowering is during August-September and fruiting in October-January. It's root-bark is used medicinally.

Medicinal Uses: Root-bark is known as a uterine tonic. It showed a toning and conctractile action on the uterus and its use is, therefore indicated before the menses, to help the uterine contractions to bring about a proper flow of menstrual blood. Its fresh juice is given to females before three days of menses and further three days after menstruation in menstrual problems. Besides, fresh-viscid juice of the root-bark is useful in the congestive and neuralgic dysmenorhoea.

Scientific name : *Acacia nilotica*
Common name : Indian gum arabic tree
Family : *Fabaceae*

It is a moderate-sized tree with a short trunk, spreading crown and feathery foliage, reaching a height of 15 m. Bark is dark-brown and much fissured. Leaves are 2.5-5.0 cm long, bipinnate with spinescent stipules; leaflets are in 10-20 pairs, long, linear and glabrous. The flowers are yellow and fragrant, in axillary globose heads. Fruits are generally solitary, 8-12 seeded, white, flat and about 7.0-15.0 cm long. Flowering is in rainy season and fruiting during cold season.

Medicinal Uses: Its bark is known as astringent and demulcent and bark powder is applied externally on ulcers. The bark-decoction is used as a gargle in sore throat and toothache and also in chronic dysentery and diarrhoea.

Scientific name : *Aegle marmelos*
Common name : Bael tree/ Bengal quince
Family : *Rutaceae*

This is a small or medium-sized, deciduous and aromatic tree, about 10 m high with straight sharp, axillary and 2.5 cm long spines. Leaves are trifoliolate with 2.5-6.0 cm long and terete petioles. The flowers are greenish-white, sweet scented, about 2.5 cm in diameter and are borne in axillary panicles. The fruits are 5.0-10.0 cm across, globose, grey or grayish-yellow and hard with orange coloured sweet pulp. Flowering is during April-May and fruiting in March-April in the next year. Its fruits, seeds, flowers, leaves, bark and roots are used.

Medicinal Uses: The unripe fruits in ayurveda, are described as acrid, bitter, oily, tasty and are difficult to digest but are appetizer, binding and are used to cure dysentery and remove pain. While the ripe ones are acrid, bitter, sweet, appetizer, febrifuge and tonic. The fresh leaf-extract of the plant is reported to reduce the period of convalescence in patients suffering from cholera or cholearic diarrhoea. The pulp of its fresh fruits after being mixed with milk when administered with cubeb powder, acts as diuretic and astringent on the mucous membrane of the generative organs, therefore, useful in gonorrhoea.

Scientific name : *Alstonia scholaris*
Common name : Devil's tree
Family : *Apocynaceae*

It is a medium-sized, evergreen tree with grey bark, whorled branches and bitter milky juice. Leaves are in whorls of 5-7 and are coriaceous. Flowers are greenish-white or greenish-yellow and fragrant, in compact umbellate- corymbose cymes. Follicles, the fruits are slender, hanging in pendulous clusters. Flowering is during October-December and fruiting in January-March. Its bark, leaves and wood are used.

Medicinal Uses: In ayurveda, its leaves are stated as acrid, anthelmintic, appetizer, bitter, galactagogue, laxative, which are used to cure asthma, heart diseases, ulcer, pains, tumours and leucoderma. The bark is also known as alternative, antidysenteric, antimalarial and astringent and is used to cure gastro-intestinal troubles and as sedative and tonic.

Scientific name : *Aphanamixis polystachya*
Common name : Amoora
Family : *Meliaceae*

It is a medium-sized evergreen tree, with a large crown of branches and grayish-brown bark, reaching a height of about 6-10 m. Leaves are 30-90 cm long and vivid green; leaflets 9-15, elliptic or ovate, acuminate and opposite with obtuse base. Flowers are white, bracteate and sub-sessile.

Its seeds and the bark are used.

Medicinal Uses: The seeds of the plant are known as anthelmintic, laxative and refrigerant, the oil obtained from its seeds is used as a liniment in rheumatism. Its bark is considered astringent and is used in the diseases of liver and the spleen, in tumour and in abdominal complaints.

Scientific name : *Aquilaria malaccensis*
Common name : Agar wood tree/ Eagle wood tree
Family : *Thymeleaceae*

This is an evergreen, large tree upto 15-25 m in height. Leaves are alternate, linear-lanceolate to lanceolate and obovate, oblong and coriaceous. Flowers are white or green and bisexual, borne in terminal umbellate cymes. Flowering is during May-June and fruiting in July-August. Wood of the tree is used medicinally.

Medicinal Uses: In ayurveda it is known, for its alternative, bitter, carminative, pungent and tonic properties. The famous Agar wood or Eagle wood of commerce is derived from the fungus-infected tree through wounds caused by *Aspergillus*, *Fusarium*, *Pencillium* and *Fungi imperfectii*. As a liniment, it is applied in various skin diseases and also in rheumatism and paralysis.

Scientific name : *Azadirachta indica*
Common name : Margose tree/ Neem tree
Family : *Meliaceae*

It is a large and handsome tree, upto 18 m high with almost a straight trunk. Leaves are imparipinnate, crowded at the ends of branches, 22.0-40.0 cm long and glabrous; leaflets are 5-15, opposite, subopposite or alternate, lanceolate or ovate-lanceolate, acuminate, serrate or dentate and glabrous. The flowers are scented and are borne in numerous axillary panicles and pedicels, which are shorter than the leaves. Drupes, the fruits are 1-seeded, oblong and greenish yellow, when ripe. Flowering is in February-April and fruiting in June-August. Its bark, fruits, seeds, flowers, twigs and gum are the parts used medicinally.

Medicinal Uses: Its bark in ayurveda, is known for its anthelmintic, astringent, bitter, pectoral and refrigerant properties, and is used to relieve *kapha* and *pitta* and also in vomiting, burning sensation near the heart, fatigue, fever, thirst, bad taste of mouth, cough and to cure ulcers and inflammation.

Scientific name : *Cedrus deodara*
Common name : Himalayan cedar
Family : *Pinaceae*

This is a tall and evergreen tree, upto 75 m height with thick and fissured bark. The wood is oily, aromatic, fine textured, fairly evergreen and straight; sapwood white and heartwood is light and yellowish-brown or brown. Leaves are dark green with wavy margins. Cones are erect, 8.0-12.5 cm long and 7.5-10.0 cm wide, obtuse and their scales are closely imbricate, broadly cuneate; upper edge thin and rounded and broader than long and deciduous, leaving the axis of the cone standing erect on the branches. Seeds are 8.0-16.0 mm with triangular wings. Cone formation starting and their maturation is during September- December. Its bark, wood, resin and the oil, obtained from the plant are used medicinally.

Medicinal Uses: The bark of the plant is known as astringent and is used in fevers, diarrhoea and dysentery. The decoction of its wood is used successfully in fever, diarrhoea, dysentery and urinary diseases in the doses of 0.7-2.8 g per day. *Oleoresin* and a dark coloured oil, *turpentine*, obtained from its wood are used as a diaphoretic agents.

Scientific name : *Citrus aurantifolia*
Common name : Lemon
Family : *Rutaceae*

This is a shrub or small tree with greenish-grey bark. Leaves are coriaceous; leaflets are elliptic-oblong or ovate-lanceolate and obtuse. The flowers often are unisexual, 2.5 cm long, white and 5-10 together in a raceme. The fruits are globular, ovoid or oblong, often mamillate at the apex and yellow when ripe with thin and tightly attached rind and yellow-green pulp, and are usually acidic with pleasant smell. Flowering is during March-April and fruiting in May-June. The fruits of the plant are used.

Medicinal Uses: The fruits of the plant, are considered antiscorbutic, antiseptic, appetizing, astringent, digestive and stomachic, and are used in liver troubles and in vomiting. Its fruit juice is used as a refrigerant drink in burning sensation and in allaying the thrust. Besides, the plant is also reported to be used against syphilis and in fractures.

Scientific name : *Crataeva nurvala*
Common name : Three-layered caper.
Family : *Capparidaceae*

It is a medium-sized, deciduous tree, about 8-10 m high with brown and bitter bark. Leaves are 3-foliate, deciduous with green upper surface and lower grey surface and aromatic, when bruised. The flowers are yellow, 5.0-7.5 cm across, borne in corymbose clusters. Berries, the fruits are 2.5-5.0 cm across, ovoid and red, when ripe. Seeds are nearly smooth and black and are embedded in yellow pulp. Flowering is in February-March and fruiting during April-May. Its stem-bark, leaves, and the root-bark are used.

Medicinal Uses: In ayurveda, its bark is known to possess antilithic, anthelmintic, detergent, expectorant, laxative and stomachic properties, which is used in the diseases of chest and blood. The bark is also used as rubefacient and its powder in urinary and kidney problems. The bark-juice is given to women after child birth. Its fresh leaves are considered rubefacient. Its bark and leaves are pounded and tied in a cloth and their hot fomentation is applied in rheumatic pain

Scientific name : *Emblica officinalis*
Common name : Indian gooseberry
Family : *Euphorbiaceae*

It is moderate-sized, deciduous tree with greenish-grey or red bark, which peel off in scales and long stripes. Leaves are pinnate, distichously closest, linear-oblong and obtuse. Flowers are greenish yellow and are borne axillary in clusters along the branchlets. Flowering is during February-May and fruiting in October-March. Its fruits, seeds, flowers, leaves, bark and the roots are used medicinally.

Medicinal Uses: Ripe fruits are known as antidiarrhoeal, astringent, antidysenteric, antiscorbutic, carminative, cooling, stomachic and tonic.

These are useful in urinary problems and also are prescribed in anaemia, jaundice and dyspepsia in combination with iron. A fermented liquor prepared from the fruits is used in jaundice, dyspepsia and coughs.

Scientific name : *Erythrina variegata*
Common name : Indian coral tree
Family : *Fabaceae*

This is a medium-sized, quick growing tree, up to 18.5 m in height, with smooth, greenish or yellowish and lustrous bark, which peels off in thin papery flakes. Stem is branched, smooth and sparsely covered with short prickles. Leaves are trifoliate; flowers arc large and coral red in colour, and borne in dense racemes. Pods, the fruits are torulose. Its leaves, bark and flowers are used.

Its leaves in ayurveda are considered anthelmintic, diuretic, emmenagogue and laxative and are used to improve appetite and to cure urinary discharges. Externally, these are for relieving the pain of joints. The bark of the plant is known to possess astringent, anthelmintic and febrifugic properties and is used in liver problems and as a collyrium in opthalmia. Besides, its flowers are used in biliousness and ear problems.

Scientific name : *Ficus benghalensis*
Common name : Banyan tree
Family : *Moraceae*

It is a large and evergreen tree, upto 30 m high and extends laterally by sending down aerial roots. The aerial roots are many and sometimes develop into supporting trunks, which assist the lateral spread of the tree indefinitely. Bark is light grey-white, smooth and about 1.25 cm in thickness. Leaves are coriaceous, ovate-elliptic with subcordate or rounded base. Male flowers are many, borne near the mouth of receptacle and are with 4 sepals. Gall flowers are with perianth and the elongated styles. Male and female flowers are found on same receptacle. The fruits are sessile in pairs, puberulous, sub-globose, 1.3-2.0 cm across and red, when ripe. Flowering is during summer and fruiting in rainy season.

Medicinal Uses: Its seeds are used as cooling and tonic. A poultice of the leaves is applied to abscesses and tender leaf-paste with honey is considered beneficial in '*raktapitta*'. The milky juice obtained from the branches of the plant, is applied externally for pains, in rheumatism and lumbago, sores and ulcers of the feet, when cracked or inflamed. This is also used in toothache and after mixing with the sesame oil, is applied on burns and in genital diseases.

Scientific name : *Ficus racemosa*
Common name : Cluster fig
Family : *Moraceae*

It is a medium-sized or large tree, with spreading branches and about 9.0-12.0 m height. The bark is smooth and reddish brown. Leaves are alternate, ovate or obovate-oblong or oblong-lanceolate. The fruits are borne in large clusters on a branches, derived from the trunk and main branches, and are subglobose. Its leaves, fruits, roots, bark and the latex are used.

Medicinal Uses: A powder of its leaves after mixing with honey, is given in bilious affections and their decoction is used in bronchitis. Its fruit is astringent, stomachic, carminative and used in menorrhoea and haemoptysis. The juice of its roots is used in dysentery and that of fruits is given as a vehicle far medicine in diabetes and urinary complaints.

Scientific name : *Ficus religiosa*
Common name : Sacred fig tree
Family : *Moraceae*

It is large or medium-sized deciduous tree with spreading branches and about 10-12 m high. Bark is grey with brownish specks, smooth and 1.3 cm thick. Trunks are irregularly shaped. Leaves are alternate, broadly ovate or rounded, caudate and somewhat pendulous, glaucous on upper surface, 5-7 veined and long petioled. Male flowers are very few and sessile: sepals 3, stamen 1 and filament is short. Female and gall flowers: sepals 5, style short and lateral. The fruits are sessile in axillary pairs and are depressed-globose, 1.3 cm across and dark purple, when ripe. Flowering is during summer and fruiting in rainy season. Its bark, leaves, fruits and the seeds are often used medicinally.

Medicinal Uses: Its bark is astringent, which is used in gonorrhoea. Its pulverized bark is applied externally on unhealthy ulcer and wounds to promote granulation and is found very efficacious when rubbed with honey to apthous sores of children. A paste of its dried bark is used as an absorbent in inflammatory ulcers. The oil medicated with its leaves is used as eardrops in earache. Its leaves and the twigs are used as laxative. Its fruits are also used as mild laxative and digestive. Besides, its seeds are also known as laxative and in powder form, are given for three days during menses as to sterilize women.

Scientific name : *Garcinia morella*
Common name : Indian gamboge tree
Family : *Clusiaceae*

This is small or medium-sized evergreen tree, up to 9.0-15.0 m in height with hard, mottled and yellow wood. Leaves are coriaceous, sagitate or ovate-lanceolate, opposite and 10.2-15.2 cm long. Flowers are unisexual, sessile and greenish-white. Flowering and fruiting is during October-April. Its stem-bark and seeds are used.

Medicinal Uses: *Gamboga*, a gum-resin, which is obtained from its stem-bark, is considered a powerful hydragogue and strong purgative. It is used in refractory constipation, as anthelmintic and in menorrhoea and ascites. Besides, its seed-butter is used as a remedy in dysentery, and mucous diarrhoea.

Scientific name : *Gmelina arborea*
Common name : Gumhar/ Malay bush beech
Family : *Verbenaceae*

It is an unarmed, large, pubescent and deciduous tree with about 20 m height and somewhat corky bark. Leaves are large, tawny-tomentose underneath when young and are cordate, ovate, acuminate. The flowers are yellow tinged with brown in terminal and axillary racemiform cymes. Inflorescence, calyx and corolla are toothed with dense soft and tawny tomentum. Drupes, the fruits are ovoid or pyriform, smooth, 1.7-2.5 cm long and orange-yellow, when ripe. Seeds are oblong. Flowering is during February-April and fruiting in April-July. Its fruits, leaves and the roots are used.

Medicinal Uses: The fruits of the plant are refrigerant and their decoction is given in fever and bilious affections. Its leaves are demulcent and their juice is useful in cough, gonorrhoea, foetid discharge and maggot in ulcers. A paste of its leaves is applied on head in headache. Its roots are bitter, tonic, demulcent, laxative, refrigerant and stomachic and their decoction is given in fever and indigestion. The root-bark is known to improve thirst and to relieve the abdominal pain. Besides, various parts of the plant, are used in swellings of throat.

Scientific name : *Helicteres isora*
Common name : East Indian screw tree
Family : *Sterculiaceae*

It is a sub-deciduous, small tree or shrub, upto 1.5-3.0 m in height. Its young branches are rough and are covered with scattered stellate hairs. Leaves are serrate, obliquely cordate or obovate, shortly acuminate and rough above and pubescent beneath. The flowers are solitary or in sparse clusters with red reflexed petals, becoming pale blue when old. The fruits are 5.0 cm long, greenish-brown, beaked and cylindrical with 5 spirally twisted carpels. Seeds are tubercled and many. Flowering is during rainy season and fruiting later in August-September. Its fruits, bark, seeds and the roots are used.

Medicinal Uses: The fruits of the plant are known as demulcent and astringent and are used in the griping of bowels and flatulence of children. Its bark is used in dysentery and diarrhoea. A powder of its seeds mixed with pure castor oil or coconut oil is prescribed in otorrhoea. Its roots are considered astringent, demulcent, diuretic and antigalactogogue and their juice or decoction is prescribed in dysentery and griping pain of the bowels, flatulence and in diabetes.

Scientific name : *Holarrhena antidysenterica*
Common name : Conessi/ Easter tree/ Ivory tree
Family : *Apocynaceae*

It is a small and deciduous tree with rough, brown bark, which exfoliates in irregular flakes. Leaves are elliptic, oblong and ovate or ovate-oblong; their nerves are 10-14 paired and strong arched. The flowers are white and slightly scented. The fruits are of two distinct and divaricate follicles, which are slender, parallel, terete, coriaceous and usually with long white spots. Seeds are numerous, 8.0 mm long, linear and glabrous. Flowering is during April-July and fruiting in August-October. Its seeds, leaves, bark and the fruits are used.

Medicinal Uses: Its bark is described as acrid, anthelmintic, bitter, dry and pungent, which is used to cure diarrhoea, fever, piles, leprosy, thirst, skin-diseases, biliousness, and the diseases of spleen. Its seeds are known as appetizer, bitter, pungent, anthelmintic and cooling and are used to cure pains, dysentery, burning sensations, bleeding piles and fatigue.

Scientific name : *Hydnocarpus kurzii*
Common name : Chaul-mogra
Family : *Flacourtiaceae*

This is an evergreen tree, upto 15 m in height with tall trunk and narrow crown of hanging branches. Leaves are thinly coriaceous, entire, 17.5-20.5 cm and lanceolate or oblong-lanceolate. Petiole is geniculate at upper end. The flowers are mostly dioecious and pale yellow, which are borne in axillary cymes. The fruits are chocolate-brown, globose and 6.5-7.5 cm across with numerous seeds. Flowering is during May-June and fruiting in November-December. The seeds and bark of the plant are used.

Medicinal Uses: An oil obtained from the seeds of the plant, which is known as '*Chaulmogra oil*', is used in leprosy and other skin disease both externally and internally. The bark of the plant is said to be used as a febrifuge and an infusion made from it, has the odour of the essential oil of bitter almonds.

Scientific name : *Hydnocarpus laurifolia*
Common name : Chaul-mogra
Family : *Flacourtiaceae*

This is a tall, evergreen and dioecious tree, upto 15 m or more in height with fluted stem and rough brown bark. Leaves are 10.5-25.5 cm long, membraneous or thinly coriaceous, oblong, ovate or elliptic and somewhat serrate. The flowers are small, white, about 1.2-2.5 cm in diameter, which are borne solitary or in fascicles. The fruits are globose, about 5.0-10.0 cm in diameter with 15-20 striate, subovoid and obtusely angular seeds. Flowering and fruiting is almost throughout the year, chiefly during January-May. Its seeds are used medicinally.

Medicinal Uses: Its seed-oil is also known by the name '*Chaulmogra oil*' and is used in leprosy and other skin disease, 15 drops gradually increased upto 30 drops of its seed-oil are recommended to be injected intra-muscularly or intravenously in the form of ethylesters or salts of chaulmogric and hydnocarpic acids in leprosy. The seed-oil mixed with limewater is also applied in sprains in rheumatic joints. Besides, a paste of the seeds mixed with sulphur, camphor, limejuice and seed oil of *Jatropha curus* is applied externally in wounds and ulcers.

Scientific name : *Madhuca longifolia*
Common name : Mahua tree, Mowra butter tree
Family : *Sapotaceae*

This is a large and deciduous tree, with about 17 m height. Bark is grey or blackish with shallow wrinkles and vertical cracks. Its young branches, leaves and petioles are pubescent or tomentose. Leaves are clustered at the ends of branches, elliptic or oblong-elliptic and shortly acuminate with cuneate base. The flowers are many at the ends of the branches and are borne on drooping pedicels. The fruits are ovoid, fleshy and green and about 2.5-5.0 cm long. Seeds are 1-4. Flowering is during April-May and fruiting in June-July. Its seeds, bark, flowers and fruits are used.

Medicinal Uses: The plant is known as acrid, aphrodisiac, anthelmintic, bitter, cooling and fattening, which cause biliousness, burning sensations, ulcers and fatigue. Its seeds are described as laxative and are used in piles and constipation. A gummy juice obtained from its seeds is applied in rheumatism and skin-infections and the seed-oil in skin diseases.

Scientific name : *Mangifera indica*
Common name : Mango
Family : *Anacardiaceae*

It is a large, evergreen tree with widely spreading branches and thick dark grey bark and about 10-40 m in height. Leaves are crowded at the ends of the branches and are alternate, simple, oblong, linear-oblong, elliptic or ovate-lanceolate, acute or acuminate. The petiole is swollen at the base and about 2.5-6.0 cm long. The flowers are greenish-yellow and odorous in terminal panicles. Flowering is during March-April and fruiting in June-July.

Medicinal Uses: The leaves of the plant are used in scorpion stings and their decoction mixed with honey is given in aphasia. The midribs of its leaves are used to remove warts on the eyelids. Its bark is known as astringent and is used in uterine haemorrhages, haemoptysis and melaena. The powder of its flowers or their decoction is useful in chronic dysentery and diarrhoea. Its ripe fruits are laxative, diuretic and astringent and are used in haemorrhage of uterus, lungs or intestines and the unripe ones are useful in opthalmia and eruptions. Besides, the powder of its seeds is prescribed in asthma, diarrhoea, dysentery, leucorrhoea, menorrhagia and bleeding piles.

Scientific name : *Melia azadirach*

Common name : Bead tree/ Persian lilac

Family : *Meliaceae*

This is a moderate-sized tree, upto 14 m height with short erect trunk and broad crown. Leaves are 22.0-40.0 cm long, bipinnate, sometimes tripinnate; leaftlets are 3-12, opposite or subopposite or alternate, ovate or lanceolate, serrate and acuminate with oblique base. The flowers are lilac, honey-scented and white, tinged with violet. Drupes, the fruits are 1.25-1.75 cm across, 5-celled and 5-seeded. Flowering is during February-March and fruiting in October-December. The fruits, seeds, flowers, leaves, bark, gum and wood-extract are used medicinally.

Medicinal Uses: Fruits are eaten as tonic and the seeds are prescribed in rheumatism. The oil obtained from seeds is applied locally in skin diseases and the paste of its flowers and that of leaves is applied in the form of a poultice to get relief in nervous headache and to cure eruptions on the scalp. Its leaf-juice is used as anthelmintic, antilithic, diuretic and emmenagogue and the decoction as astringent and stomachic.

Scientific name : *Mesua ferrea*
Common name : Iron wood/ Nahor
Family : *Clusiaceae*

It is a medium-sized or large evergreen tree with short trunk, which often buttresses at the base and the dark-red heartwood. Leaves are coriaceous and lanceolate; upper side shining and lower covered with a white waxy powder. Flowers are white, large and solitary or in clusters and fragrant. The fruits are ovoid, 2.5-5.0 cm long with persistent calyx. Seeds are 1-4, shining, dark brown with oily and fleshy cotyledons. Flowers in February-April and fruiting is during September-October. The flowers, seeds, fruits, leaves and bark of the plant are used medicinally.

Medicinal Uses: Its flowers and stamens form a major drug *Nagapushpa* or *Nagkesar* in indigenous system of medicine, is described as alexipharmic, digestive and dry. This is used as astringent, stomachic and expectorant.

Scientific name : *Oroxylum indicum*
Common name : Indian trumpet tree
Family : *Bignoniaceae*

It is a small, deciduous and soft-wooded tree with few branches and a small open crown. The bark is soft, light, brownish-grey and corky outside. Leaves are opposite, 3-pinnate near base, 2-pinnate about the middle and simply pinnate towards apex. The flowers are white or purplish and many in large and erect racemes. Capsules, the fruits are flat and straight. Seeds are many. Its root-bark, stem-bark, fruits, seeds and leaves are used medicinally.

Medicinal Uses: Its root-bark is considered astringent and tonic and is used in diarrhoea and dysentery. After boiling in oil, it is used in otorrhoea and its powder along with '*haldi*' is used to cure the sore-backs and its infusion in rheumatism and as bitter tonic. Its tender fruits are used as refreshing and stomachic and the seeds as purgative. Besides, the decoction of its leaves is said to be given in stomach-ache and externally to be applied for spleen enlargement.

Scientific name : *Phoenix sylvestris*
Common name : Date sugar palm/ Wild date palm
Family : *Arecaceae*

It is a tall and erect palm, reaching a height upto 18 m and with a thick hemispherical crown. Trunk is solitary and clothed with the persistent bases of the petioles. Leaves are pinnate, grayish-green, 2.1-3.6 m long and glabrous. Berries, the fruits are oblong-ellipsoid, orange-yellow to reddish-brown and edible. Seeds are rounded at the ends, grooved to centre on one face, pale brown and are about 1.6 cm long. Flowering is in summer and fruiting later during September-October. Its fruits and roots are used.

Medicinal Uses: Fruits are tonic and restorative, hence these are advised to be eaten raw. The fresh juice obtained from its stem is used as a cooling beverage, the sap of trunk is cooling, slightly narcotic and digestive and it is prescribed for diabetic persons.

Scientific name : *Pterocarpus santalinus*
Common name : Red sandalwood tree
Family : *Fabaceae*

It is a small-sized, deciduous tree, attaining a height of 11 m with extremely hard and dark purple heart wood. Leaves are generally imparipinnate. Flowers are yellow, borne in simple or sparingly branched racemes. Pods, the fruits are woolly and about 3.8 cm across. Seeds are smooth, coriaceous and reddish brown. Flowering and fruiting is during summer season.

Medicinal Uses : The wood of the plant is known to possess astringent and cooling properties and is used beneficially in fever, dysentery and haemorrhage. Externally, a paste of its wood is applied on forehead in headache, in inflammation, boils and piles and also in ophthalmia.

Scientific name : *Salix caprea*
Common name : Goat willow/ Sallow
Family : *Salicaceae*

This is a small tree. Leaves are 5.0-10.0 cm long, ultimately glabrous and more or less rugose above, grey-tomtentose beneath and crenated or nearly entire. Stipules are usually conspicuous and sub-reniform. Flowers are appeared before the leaves and are sweet-scented. Capsules, the fruits are downy and stiptate. Flowering and fruiting is during September-December

Medicinal Uses: The plant is cooling, bitter and sharp tasty, and is used in pain and inflammation of the liver, thirst, biliousness, headache and as aphrodisiac. A distilled water is prepared from its flowers, which is used as cordial, stimulant and aphrodisiac. Externally, it is applied in headache and in ophthalmia. The ashes of its wood are applied in haemoptysis and after mixing with vinegar are applied in haemorrhoids.

Scientific name : *Santalum album*
Common name : Sandal tree
Family : *Santalaceae*

It is a small, evergreen, glabrous and semi-parasitic tree with slender branches, attaining a height upto 18 m with dark grey or nearly black or reddish and rough bark. Sapwood is unscented and white but heartwood is scented and yellowish-brown or dark-brown. Leaves are opposite, ovate or ovate-lanceolate, glabrous, larger and thin. The flowers are brownish purple, violet or straw-coloured, unscented and are borne in terminal and axillary paniculate cymes. Drupes, the fruits are globose, 1.2 cm across, and purple black with hard ribbed endocarp. Seeds are ovoid or globose. Flowering is during June-September and fruiting in November-February. Its wood and sandalwood oil are used.

Medicinal Uses: A paste of its wood ground up with water, is applied to the temples in headache, fevers and local inflammations and to skin diseases, to allay headache and also used as diaphoretic. A decoction made out by mixing its wood with dried ginger, is successfully used in haemorrhoids. Sandalwood oil and oil derived from its heartwood is considered cooling, diuretic, diaphoretic and expectorant.

Scientific name : *Semescarpus anacardium*
Common name : Marking nut tree
Family : *Anacardiaceae*

It is a medium-sized, dioecious and deciduous tree with rough and dark-coloured bark. Its young branches, inflorescences and undersides of the leaves are crowded towards the extremities of the branches and are large, oblong-obovate, rounded at apex, cartilagious at margin and coriaceous. Flowers are greenish-yellow, fasciculated and are arranged in erect, compound and terminal panicles. Drupes, the fruits are about 2.5 cm long, obliquely oval or oblong, smooth, shining and purplish-black and are seated in an orange cup. Flowering is during May-June and fruiting in November-February. Its fruits and bark are used medicinally.

Medicinal Uses: Fruits are used in asthma, ascites, epilepsy, neuralgia, psoriasis and rheumatism and also as abortifacient and vermifuge. A decoction of its fruits mixed with milk and butterfat is used efficaciously in asthma, gout, hemiplegia, sciatica and syphilis. After bruising, the fruits are applied to *os uteri* to procure abortion and also given as vermifuge. An oil obtained from its nuts is used as vesicant and externally in rheumatism and leprous nodules.

Scientific name : *Stereospermum chelonoids*

Common name : Trumpet flower

Family : *Bignoniaceae*

It is a large, deciduous tree, about 9-18 m high with dark-grey bark, which exfoliates in large and flat scales. Leaves are opposite, about 30-60 cm long and paripinnate. The flowers are dark purple coloured, pubescent and fragrant, in viscid and trichotomous panicles. Capsules, the fruits are loculicidally 2-valved, 4-ribbed and dark grey with white specks. Its root-bark, seeds and flowers are used.

Medicinal Uses: The root-bark of the plant is considered cooling, diuretic and tonic and its decoction is used to remove the calculi. A paste of its seeds is applied on forehead in migraine. Besides, its flowers are rubbed up with honey and are given to check the hiccough and as an aphrodisiac.

Scientific name : *Strychnos nux-vomica*

Common name : Strychnine tree/ Nux vomica

Family : *Loganiaceae*

It is an evergreen and glabrous tree, often with short, sharp, strong and axillary spines and thin grey, smooth or rough bark, reaching a height upto 30 m. Leaves are ovate, 5-nerved, glabrous, shining with obtuse base. Flowers are greenish-white, and borne in terminal, pubescent, pedunculate and corymbose cymes. Flowering is during March-April and fruiting in winter season. Its root-bark, fruits, leaves, wood and seeds are used.

Medicinal Uses: The fruits of the plant are said to be acrid, appetizer, bitter, pungent and tonic and are used to cure leucoderma, '*vata*' and '*kapha*' diseases, diseases of the blood, anaemia, piles, jaundice and urinary dischargement. In Unani system, these are described as aphrodisiac, bitter, diuretic, emmenagogue and tonic and are used to cure pain and weakness of joints, lumbago, piles and paralyis.

Scientific name : *Symplocos racemosa*
Common name : Lodh tree
Family : *Symplocaceae*

It is a small, glabrous and evergreen tree with trunk upto 6.0 m high and 15.0 cm in diameter. Leaves are coriaceous, elliptic-oblong, serrulate with glabrous or pilose midribs. Flowers are yellow, fragrant, nearly sessile and are borne on short axillary and compound spikes. Drupes, the fruits are oblong or cylindric, 1.0-1.2 cm long, more or less distinctly ribbed, often slightly curved and purple coloured, when ripe. Flowering is during October-January and fruiting in December-May. The bark and wood of the plant are used medicinally.

Medicinal Uses: The bark of the plant in ayurveda, is known for its acrid, astringent, cooling and digestible properties and is used in eye diseases, spongy gums, bleeding and to treat '*kapha*' diseases.

Scientific name : *Syzygium cumini*
Common name : Jambolan/ Java plum
Family : *Myrtaceae*

It is a large and evergreen tree with light grey bark. Leaves are coriaceous, shining, entire, oval-oblong or lanceolate-oblong, long-acuminate, tough and smooth, lateral nerves are numerous, parallel and confluent near the margins. The flowers are tetramerous, small, greenish and sessile and are arranged in broad trichotomous panicles. Berries, the fruits are oblong-obovoid or subglobose, pink and turning black, when ripe and are smooth and shining. Flowering is in March-May and fruiting during May-July. The fruits, seeds, leaves and bark are used.

Medicinal Uses: In ayurveda, its fruits, are described as acrid astringent, cooling, dry and sweet, which can remove bad smell from the mouth and biliousness and are good for diabetes. Its seeds are hypoglycaemic and are used to allay the thirst in diabetes. Its leaf-juice is used in dysentery. The powder of its bark is used as gargle and as washes in leucorrhoea, ringworm and skin diseases.

Scientific name : *Tecomella undulata*
Common name : Rohida tree/ Wave-leaved Bignonia
Family : *Bignoniaceae*

The flowers are dark yellow in corymbose racemes, which terminate into short lateral branches. Flowering is during April-May and fruiting later in June-July.

Medicinal Uses: In ayurveda, its bark is said to be used in urinary discharges due to *kapha* and *pitta*, the enlargement of spleen, and leucoderma. The bark of young branches of the plant, is used as a remedy for syphilis and leucorrhoea. Moreover, the extract of its stem-bark is used as relaxant and chloretic.

Scientific name : *Terminalia bellirica*
Common name : Belleric myroblan/ Vibhitaka
Family : *Combretaceae*

It is a large and deciduous tree with bluish or ashy-grey bark and about 20-25 m in height. Leaves are clustered at the ends of the branchlets and are coriaceous, pale beneath, broad elliptic or obovate-elliptic, 8.0-20.0 cm long and often with unequal base. Petiole is 2.5-7.5 cm long and usually glandular. Spikes are slender, interrupted and 8.0-15.0 cm long. Flowers are about 1.25 cm across, dirty-grey or greenish-yellow with offensive smell. The fruits are 2.5 cm long, ovoid, ellipsoidal or globose and grey velvety with 5, more or less indistinct furrows. Nuts are thick walled and hard. Flowering is during April-June and fruiting in December-January.

Medicinal Uses: Fruit bitter, astringent, tonic, laxative, antipyretic and antidiarrhoeal, which are used in piles, dyspepsia, leprosy, biliousness, headache, hoarseness and eye diseases. Its ripe fruits are used as an astringent in combination with chebulic myrobalan and half-ripen ones as purgatives.

Scientific name : *Terminalia arjuna*
Common name : Arjun tree
Family : *Combretaceae*

It is a large tree, often with buttressed trunk, smooth grey bark and drooping branchlets and about 20-25 m in height. Leaves are usually sub-opposite, oblong or elliptic-long, pale-dull green above and pale-brown beneath, 10.0-20.0 cm long and hard. The flowers are yellowish-white and are borne in shortly panicled-spikes. Flowering is in April-May and fruiting during September-Novemeber. Its fruits, leaves and the bark are used.

Medicinal Uses: The fruits of the plant are used as tonic and deobstruent. Externally, its leaves are used as a cover on sores and ulcers and their juice is dropped in earache. The bark is antidysenteric, antipyretic, astringent, cardiotonic, styptic and tonic. Powder of the bark acts as a diuretic in cirrhosis of liver and gives relief in symptomatic hypertension. A decoction of its thick bark made with milk and *gur* is used as a cardiotonic.

Coniferous trees

Scientific name : *Abies alba*

Common name : Silver fir

Family : *Pinaceae*

From the Latin *abire*, to go away, used here in the sense of distance from the ground to refer to the height that some species attain. *Alba* means white. Grows in the mountains of central and southern Europe. A tree up to 60m tall with a straight, columnar trunk and smooth, ashy-white bark with resin sacs when young, later cracking and secreting resin.

Leaves: The leaves grow in two opposite rows; they are 2-3 cm long, dark green but rather lighter above, with two silver-white stomatic lines on the underside which give the tree its name, Silver Fir.

Cones: The cones are erect and greenish brown, the scales have a characteristic outgrowth and fall when ripe, leaving the axis bare on the tree.

Propagation: By seed, shoots can only be obtained if treated with rooting hormones (auxin).

Scientific name : *Agathis australis*

Common name : Kauri pine

Family : *Araucariaceae*

From the Greek *agathis*, a ball of twine, since the overlapping scales on the cones look altogether like rolled-up threads. Native to New Zealand. An evergreen tree which at first sight might seem to belong to the broadleaves rather than the gymnosperms, its straight, gray trunk grows to heights of 90m and more.

Leaves: The leaves are linear-oblong, rosy or bronzy in colour; when mature they become oval and bright green, up to 5cm in length and well spaced out on secondary twigs.

Flower: The unisexual flowers grow at the axil of the sessile leaves, the males in solitary catkins, while the females are strobiles, almost spherical, 8 cm in diameter, with a single large seed at the base of each scale.

Propagation : By seed or by cuttings of young branches, taking them with a heel if possible.

Scientific name : *Araucaria araucana*

Common name : Monkey's puzzle tree, Chile pine

Family : *Araucariaceae*

The tree is named after a South American Indian tribe, the Araucana, who live in the regions where it grows wild, Western slopes of the Andes in Chile. A pyramidal evergreen tree up to 30 m high, with whorled branches, generally five to a node, growing horizontally near the trunk but upturned towards their ends.

Leaves: The leaves about 5 cm long, are rigid, ovatoacuminate, closely overlapping so as to cover the branches completely.

Propagation: By seed, under glass at the end of winter. For fertilization to take place it is necessary to have trees of both sexes growing close together; they seldom fruit in our climate.

Scientific name : *Araucaria bidwillii*
Common name : Bunya-bunya
Family : *Araucariaceae*

The specific name commemorates the British naturalist John C. Bidwill (1815-1853). Native to Queensland in Australia, Where it is called *bunya-bunya*. This conifer, shaped like an inflated cone, can grow up to 50 m tall, but in mild, humid climates, which suit it well for cultivation, it assumes a more massive shape.

Leaves: The leaves are bright green, some 2 cm long, lanceolate-acuminate, strongly scented at the tips and with parallel, flattened veins.

Flowers: The flowers are unisexual, the males growing as catkins, the females in round strobiles as much as 25 cm long which, once fertilization is completed, bear a big winged seed on each scale.

Propagation: By seed; by cuttings from young branches or by basal suckers.

Scientific name : *Araucaria excelsa*
Common name : Norfolk island pine
Family : *Araucariaceae*

The specific name comes from the Latin *excelsus*, great, referring both to its splendid shape and to the height it can reach in its natural surroundings. Endemic on Norfolk Island, east of Australia. These plants, are grown in pots as house plants and look like little fir trees. It is a huge tree in its native habitat, growing up to 60 m, with erect habit and a thick, strong trunk whose branches grow horizontally, regularly whorled, so that the crown becomes pyramidal.

Leaves: The leaves appear in two shapes; those on the youngest branches are light green and unscented, while those on older branches are shorter and overlapping, with a stiff point.

Flowers: The flowers are unisexual, and the females produce rounded strobiles some 10 cm long and 11 cm broad, with a large, winged seed to every scale.

Propagation: By seed, but since growth is very slow, propagation by terminal cuttings is generally preferred. The lateral shoots that grow where the cuttings are taken can be used for further cuttings.

Scientific name : *Cedrus atlantica*
Common name : Atlas cedar
Family : *Pinaceae*

The name is derived from the Latin *cedrus* and Greek *kedros*, a tree that has not been definitely identified but was probably a species of *Juniperus*. Native to the mountainous region of Algeria and Morocco, but some writers mention that in more remote periods of the earth's history the cedar grew wild in Europe. The tree reaches a height of 50 m with a girth of 1.5 m. It has a crown with a completely erect summit.

Leaves: The leaves are needlelike, short and stiff, flattened and curved, glossy green, joined in little tufts.

Cone: The cones are short and erect with a depression at the tip, at first yellowish, then violet-purple; they contain seeds with wings 2 cm long.

Propagation: Only by seed, which must be ripe.

Scientific name : *Cedrus deodarus*
Common name : Deodar
Family : *Pinaceae*

The species name is derived from the Sanskrit *devadara*, tree of the gods (hence the common Indian name, *deodar*). Native to the Himalayas and Baluchistan, at 1100-4000 m. Grows up to 50 m high; the crown is pyramidal, with an inclined head when young.

Leaves: The leaves are needlelike, 2.5 - 5 cm long and smooth.

Cones: The cones, 7-12 cm long, are conical, rounded at the end with no depression, violet at first turning to brown, and covered with smooth-backed scales; they ripen biennially.

Propagation: Like other cedars, by seed.

Scientific name : *Cupressus macrocarpa*
Common name : Monterey cypress
Family : *Cupressaceae*

The species name refers to the large size of the fruit. A tree that can grow to a height of 25 m and has a thick, dark brownish-red bark, turning light gray in old specimens and breaking off in flat scales.

Leaves: The leaves, scale like and lying flat against the branches, light green, and are triangular and aromatic.

Strobiles: The strobiles are formed of 10-12 polygonal scales, angular and pointed; they are green at first, then gray, brown, a rather light greenish gray-brown, and finally brownish-violet.

Propagation: By seed, grows very quickly.

Scientific name : *Cupressus sempervirens*
Common name : Italian or Mediterranean cypress
Family : *Cupressaceae*

The specific name marks the persistence of the foliage, giving the crown its dark-green colour. Grows wild in regions bordering the eastern Mediterranean. A tree 20-30 m high, with a great variety of forms of crown, among the most typical being the spreading form of the var.*horizontalis* and the pyramidal var.*stricta*, wrongly called female and male cypress.

Leaves: The scalelike leaves, growing along the branchlets like tiles, are dark green with resinous glands which secrete *oleum cupressi*, used in the pharmaceutical industry.

Cones: The rounded cones, 3-4 cm in diameter, are composed of 8-12 scutate woody scales.

Propagation: By seed.

Scientific name : *Cycas revoluta*
Common name : Sago palm
Family : *Cycadaceae*

The name comes from the Greek *kykas*, used by Theophrastus for a plant not definitely identified. Originally from eastern Asia, from southern Japan to Java. Though very like a palm, this slow growing little tree will reach a height of only 3.5 m in ideal conditions.

Leaves: The leaves are as much as 2 m long, deflexed and pinnate, with a number of thin, stiff segments with sharply pointed tips, forming a big crown at the top of the trunk.

Flowers: The trees are dioecious, with very simple flowers; the males resemble strobiles made of scales and carry the pollen sacs; the female grow in a big rosette of carpellary leaves with the ovules on the lower part, protected by a thick crown.

Propagation: By seed or basal suckers.

Scientific name : *Gingko biloba*
Common name : Maidenhair tree
Family : *Gingkoaceae*

The name comes from the Japanese ginkyo, used in the 17th century and wrongly transcribed. Many modern authorities have adopted the spelling 'Ginkyo'. Native to China, where fossils of this plant have been discovered dating from the Mesozoic period. The tree has been able to survive because it was cultivated by monks. A deciduous tree, dioecious, which reaches a height of 40 m; the bark is reddish.

Leaves : The flatbellate leaves have the upper margin irregularly toothed, and are divided by a deep incision which cuts them into two lobes; they have a great number of veins, and are glossy green in colour, though becoming golden yellow before they fall.

Cones: The reproductive elements are insignificant; the male flowers grow as cartkins, the stalked females spring from the axils of scale-like bracts and are followed by fruit in the form of false drupes with a fleshy, evil-smelling pericarp.

Propagation: By seed, or by cuttings of soft wood, or by grafting in order to avoid female specimens. The tree will take 30 years to reach a height of 10 m.

Scientific name : *Juniperus communis*
Common name : Common juniper
Family : *Cupressaceae*

The name has kept its original Latin form. A species widely distributed in almost the whole of the Northern hemisphere, from sea level to 3750 m. Generally bushy or trailing, the plant has aboreal forms which can reach a height of 10 m though the trunk is never more than 10 cm in diameter.

Leaves: The needle like aromatic leaves, 10-14 mm long, are glaucous green.

Flowers: The globose female flowers have a number of sterile scales underneath and three fertile scales above; these last, after fertilization, swell up and grow together to form a round, fleshy body, wrongly called a berry strictly, called a cone.

Propagation: By seed; a slow grower

Scientific name : *Pinus cembra*
Common name : Cembran pine, Swiss stone pine
Family : *Pinaceae*

Pinus from the Latin, *cembra* from its Italian name. Distribution is divided into two sectors: in Europe, in the Alps and the Carpathians, and in Asia, from the Urals to northern Japan.

Leaves: This is the only European conifer to have the leaves triangular in section; they are rather stiff, dark greenish and glaucous, growing in clusters of five.

Propagation: By seed.

Scientific name : *Pinus excelsa*
Common name : Himalayan pine, Bhutan pine
Family : *Pinaceae*

The species name refers to its elegant appearance. It is also called *P. wallichiana*, after Nathaniel Wallich, or *P. griffithii* after William Griffith.

Seen at altitudes from 1600 to 4000 m in the southeast Himalayas as far as Afghanistan; in Burma and Yunnan. Sometimes called the Weeping Himalayan Pine on account of its long leaves (12-15 cm long), partly pendant and silver-blue and its long, pedunculate, pendulous cones, are ornamental.

Propagation: By seed

Scientific name : *Pinus halepensis*
Common name : Aleppo pine
Family : *Pinaceae*

The tree gets its specific name from the ancient city in Syria. Grows in the Mediterranean region, where it has become widespread as a result of cultivation. Not a tall tree, its trunk is often twisted. It may live upto 150-200 years.

Leaves: The needlelike leaves are joined in pairs, or more rarely in threes.

Cones: The cones, ovate-conical, contain small seeds.

Propagation: By seed sown directly in the ground or in seedbeds in autumn.

Scientific name : *Pinus Nigra*
Common name : Austrian pine/Black pine
Family : *Pinaceae*

The specific name refers to the colour of the crown. The species is distributed in quite small isolated areas, from Spain to the Crimea, from Asia Minor to Austria and from Algeria to Morocco. The Black Pine is a tall tree, as much as 40 m, very long lived and resinous, generally with a straight, dense crown and whorled branches growing horizontally.

Leaves: The linear leaves are pointed, 8-18 cm long, in pairs; they are dark green and very glossy

Flowers: Male and female flowers

Cones: The ovate-conical strobiles, solitary or in groups of two or four, are brightly coloured.

Propagation: By seed.

Scientific name : *Podocarpus neriifolius*
Common name : Mounteak, Thitmin
Family : *Podocarpaceae*

The name comes from the Greek *podos*, foot and *karpos*, fruit from the fruit's fleshy stalk. Native to the Himalayan region, but with varieties as far as in the Sunda Islands. Large, evergreen tree which can grow to a height of 15-20 m, with a solid, much branching trunk and a spreading crown.

Leaves: The leaves are sparse, sometimes whorled but always rather close growing, especially on the secondary branchlets; they are lanceolate-acuminate, narrower at the base, with a short stalk and prominent central veins, glossy dark green, not uncommonly lighter or translucent towards the edge; on maturity they may become slightly curled, and by that time they reach a length of more than 10 cm

Flowers: The flowers are unisexual, nearly always solitary. The males are sometimes clustered in brown catkins; the females, greenish, with a succulent stalk, give way to a naked seed which ripens in about two years, with a fleshy, ovoid, greenish growth on a dark stalk about 1 cm in diameter.

Propagation: By seed.

Scientific name : *Sequoiadendron giganteum*
Common name : Wellingtonia/Giant sequoia/"Big tree"
Family : *Pinaceae*

The name comes from that of the genus *Sequoia*, to which the plant is often assigned, and from the Greek *dendron*, tree. Distributed in a limited area in Central California, these are among the tallest and most long lived trees in the world; the trunk can reach a height of 96 m and the age of many trees may be anything between 2000 and 3500 years, though it is not possible to be certain since no specimen has ever died of old age - only as the result of some accident.

Leaves: The buds grow from the persistent leaves, which are scalelike and pointed, up to 6 mm long.

Flowers: The male flowers form axillary and terminal clusters, the female, apical clusters made up of whorled scales each having 4-7 ovules at the base.

Propagation: By seed.

Scientific name : *Thuja occidentalis*
Common name : American Arbor vitae/White cedar
Family : *Cupressaceae*

The name seems to be derived from the Greek *thyon* or *thia*, a tree that produced resin or incense; which was burnt as incense during religious ceremonies. This species lives in southeastern Canada and the northeastern United States, from Nova Scotia to North Carolina. A tall tree, generally 18-20 m, with a pyramidal crown when young, becoming irregular later; thick trunk divided in two or three at the base, with horizontal branches curved at the ends and pendulous branchlets.

Leaves: The leaves, opaque green on the upper surface and yellowish green below, are small and scalelike, overlapping and persistent.

Strobiles: The strobiles are ovoid, often growing in groups; they are composed of 5-6 pairs of scales, not pointed, only two pairs of which are fertile; the seeds are winged.

Propagation: By seed; also from cuttings.

Scientific name : *Thuja orientalis*
Common name : Oriental arborvitae/Chinese thuja
Family : *Cupressaceae*

The proper name is *Biota orientalis*, from the Greek *bios, life;* it is not certain whether it was given this name for its capacity to tolerate unlimited pruning or because it was used as a cure for scurvy. Native to Manchuria and Korea, but cultivated in all Asian countries, of modest size, not more than 12 m tall. Both branches and shoots are erect and the former are often no longer than the diameter of the trunk.

Leaves: The bark is thin and red-brown; the leaves persistent, scale like and opposite, are smaller than those of other species and have no glands.

Strobiles: The strobiles, ovoid and fleshy, glaucous green at first turning reddish brown, are formed of 6-8 thick scales that bear downward hooks which protect the 1-2 wingless seeds.

Propagation: By seed and by cuttings

Scientific name : *Torreya californica*

Common name : California nutmeg

Family : *Taxaceae*

Named in honor of the American botanist John Torrey (1796 - 1873). Native to California.

A large evergreen tree, often more than 20 m tall; the bark is gray-brown and the branching fairly open; the secondary branches are pendulous.

Leaves: The linear leaves up to 8 cm long, grow spirally round the branches, but because the short stalks are twisted it looks as if they were opposite; they are glossy dark green on the upper side, pale with two yellow ridges on the underside, and they give off an unpleasant smell if rubbed. The flowers are unisexual, the males are joined so as to form little axillary strobiles, the females consist of a single ovule, with a scaly base.

The fruits are long oval drupes and consist of one seed within a hard shell surrounded by a fleshy aril, light green with violet-red stripes.

Propagation: By seed, by softwood cuttings and by grafting.

Palms

Scientific name : *Areca catechu*
Common name : Betel palm
Family : *Arecaceae*

A slender palm with erect, rather thin stem, solitary, in its homelands reaching 10-30 m in height. It has a crown of big fronds up to 2 m long, pinnate with fairly large segments, dark green, glabrous and irregularly toothed at the tip, with unprotected stalk. Some of the outer leaves are pendulous, but the general habit is erect. The flowers are unisexual, growing in an inflorescence with branching spadix that appears below the fronds, often apart from them. The female flowers grow singly, at the base of the ramification, surrounded by the smaller, white scented male flowers. The fruit is an ovoid berry, about 5 cm long and contains a dye stuff. These seeds, boiled, sliced and dried and wrapped in the leaves of *Piper betle*, are chewed by all East and South Indian people for their slight inebriating effect.

Scientific name : *Ceroxylon andicola*
Common Name : Andes palm
Family : *Arecaceae*

The very tall stem, up to 30-40 m high and 1-2 m in circumference, marked in rings by the scars left by fallen leaves, is covered with a coat of gray wax which oozes out, especially from the leaf nodes, and flows over the trunk, making it look like a marble column. The leaves up to 4 m long, are pinnate with acuminate segments, dark green on top and whitish underneath; the inflorescence is a spadix that forms a beautiful yellow cluster. The fruits are single-seeded berries as big as billiard balls, which turn reddish purple when ripe. The wax produced by this palm has local economic importance.

Scientific name : *Cocos nucifera*
Common Name : Coconut palm
Family : *Arecaceae*

Widespread and cultivated for centuries throughout the tropics, but believed to be of Indo-Malayan origin.

This palm has a relatively slender, flexible stem which can grow to a height of 30-40 m marked in rings each year by the scars of fallen leaves and crowned by an apical cluster of huge long pinnate leaves. The flowers united in inflorescences with a spadix and bracted, are unisexual with the females below, generally solitary and formed by six twining sepals, while the numerous males grow at the top of each branch. The fruit is the well-known coconut, actually a big containing a sweet liquid which becomes fleshy on ripening and when dried constitutes the copra used as a source of oil and fats. The whole plant is very important economically, both for its fruit and its wood; the shoots, too, are used for food.

Scientific name : *Cyrtostachys renda*
Common name : Sealing wax palm
Family : *Arecaceae*

The name is derived from the Greek *kyrtos* curved and *stachys*, ear or grain, in reference to the curved inflorescence.

This palm has a bare stem, erect and slender, which can reach a height of as much as 10 m. The leaves, forming a big crown at the top of the stem, are pinnate, with linear or sword-shaped segments, generally obtuse at the tip, sometimes bifid; they are grayish on the underside, and the midrib and the bases of the leaves are red. The spadix may be more than 1 m long, with roughly alternate branches reaching about 40 cm long. The flowers are unisexual, enclosed within the spadix, and the fruits small and ovoid; the stigma persisting at their tip.

Scientific name : *Elaeis guineensis*
Common name : Oil palm
Family : *Arecaceae*

The stout trunk, ringed by the scars formed by fallen leaves, may grow to 20 m and is crowned by a big apical cluster of leaves 2-5 m long, with spiny stalk and a great many linear lanceolate, acuminate segments; the outer leaves hang down, the inner ones are erect, so that the crown appears almost globose. The flowers are unisexual, carried in separate inflorescence; the males have a spiny tip, the females have spadices with rather big stalks, also spiny tipped, and appressate, so that when they ripen they form a dense, round mass of fruit with the spines sticking out of it. There may be as many as 200-300 fruits, drupes 4 cm long, with a red or blackish fleshy aril surrounding a very hard black nut with a white, soft kernel. An oil extracted from the fruit is used in the manufacture of foodstuffs and soap, and the whitish butter from the kernel is used locally in cooking.

Scientific name : *Erythea armata*
Common name : Blue palm
Family : *Arecaceae*

These are big, bare-stemmed trees terminating in a cluster of round leaves subdivided into 40-50 segments of silvery blue-green. The stalk has powerful thorns which earn the plant the epithet *armata* in the specific name. The spadices are elongated and the flowers bisexual and sessile, grow in groups of three. The fruit is more or less round, resembling that of *E.edulis*, though it is not edible. An elegant, decorative palm, it is as hardy as it is beautiful, so that it can be grown for ornamental purposes in the warmer zones of temperate climates. This attractive American palm was introduced into European gardens at the end of the 19th century, immediately after it was discovered.

Scientific name : *Euterpe edulis*
Common name : Assai palm
Family : *Arecaceae*

Supple and slender stem up to 20-30 m high, with a feathery crown of drooping leaves, often solitary but also quite often in dense groups. The leaves are sheathed, bare, with long stalks and as many as 60-80 linear segments on either side of the midrib; they are pendulous, up to 3 m long. The little flowers, white and sessile, are borne in the interior of the branching spadices; the little pealike fruits are purple. The seeds are crushed in water to produce a beverage drunk locally under the name of *assai*, and the plant is very often commonly called the Assai palm. This plant belongs to the group known as Cabbage Palms, the young shoots of which may be eaten, either cooked or raw in salads.

Scientific name : *Howea fosteriana*

Common name : Kentia/Curly palm

Family : *Arecaceae*

These elegant plants are undoubtedly among the best known of all palms. Young specimens are often grown as ornamental house plants, but in their original home, and if cultivated in congenial climates, they can reach heights of 20 m, with a stout stem marked in rings by the scars from fallen leaves. The crown, at the apex, is formed of long leaves which gradually grow bigger as the plant reaches maturity; they are borne on long, thin stalks and are pinnate, with fairly broad, leathery segments. The flowers are enclosed in spadices which appear among the bases of the opposite leaves, and are unisexual; the males have rounded sepals. The fruits are drupes, rather like olives, greenish yellow. They appear in great racemes which ripen some six years after the first flush.

Scientific name : *Hyphaene thebaica*

Common name : Doum palm

Family : *Arecaceae*

This is virtually the only palm with a branching stem; it can reach a height of 15 m. The branches are dichotomic, continuously bifurcated, each branch bearing a rosette of fanshaped leaves at the tip, not very big, spiny stalked, with pointed segments; the outer leaves are markedly deflexed and pendulous. The stems are thin and smooth, partly covered by the bases of fallen leaves; the general habit is quite different from that of other palms. The flowers are unisexual, and the fruits that follow them are orange-yellow drupes, scented and edible.

Scientific name : *Jubaea chilensis*
Common name : Chilean wine palm
Family : *Arecaceae*

This genus contains only one species, also known as *J. spectabilis.* This great palm grows as tall as 25 cm, with a stout stem covered by the bases of withered leaves and carrying at its head a great crown of long pinnate leaves with linear-lanceolate segments, rigid and forming sheathes at the base. The inner leaves are erect, the outer deflexed. The inflorescence have spadices, with unisexual flowers followed by fruit like little coconuts. The special feature of this plant is that, if the trunk is incised, it exudes a sugary liquid which will continue to flow as long as the cut is kept fresh; as much as 400 litres of syrup can be obtained in three months from an adult tree. In the tree's homeland the sap is boiled and a syrup obtained from it, called "palm honey", which can be fermented to produce an alcoholic drink.

Scientific name : *Livistona australis*
Common name : Australian fan palm
Family : *Arecaceae*

A slender palm which can reach a height of 20 m or more. The stem is marked in rings by the scars left by the bases of fallen leaves, while in the juvenile stage it is covered by the bases themselves and by brownish fibres. The huge palmate leaves, leathery and smooth, which crown the top are almost circular and more than 1 m in diameter; they are divided into two distinct parts, with narrow curved green segments with yellowish central veins and a pointed tip that may be entire or divided into two. There are big, curved thorns on the stalk. The flowers are joined in long erect spadices which become pendulous when, after fertilization, the oblong yellow or brownish fruits appear.

Scientific name : *Metroxylon sago*
Common name : Sago palm
Family : *Arecaceae*

These palms are often shrubby, with several stems which reach a height of about 12 m and a diameter of 70 cm. The large, suberect pinnate leaves with linear, acuminate, opposite segments, can also be as long as 6 m. The flowers are polygamous or monoecious, carried by long, branched spadices which form a sort of broadened spike. The fruits are globose or elliptical drupes some 3 cm long, with a single seed enclosed in overlapping scales; they take several years to ripen. Each stem dies after fruiting, but the stump puts out new shoots. The pith is extracted by felling the tree before it fruits, cutting into strips and washing it out. The starch remaining is the sago, which is used for food.

Scientific name : *Phoenix canariensis*
Common name : Canary islands date palm
Family : *Arecaceae*

Perhaps the best known of the ornamental palms, since it can be grown in pots when young, while in the Mediterranean countries the adult trees make a beautiful show in gardens. This is a vigorous tree that can reach a height of 20 m. The leaves are pinnate, 5-7 m long, with longitudinally acuminate segments, sometimes transformed at the base into long double thorns at the node. The spadices are borne on long stalks, 1 m or more long, and curved; the fruits, about 2 cm long, are ovoid or spherical, with a yellow skin turning reddish. The tree differs from *P.dactylifera* in having a shorter, thicker trunk and wider, stiffer leaf segments; these are bright green.

Scientific name : *Phoenix dactylifera*
Common name : Date palm
Family : *Arecaceae*

This palm, whose stem, covered with the bases of withered leaves, can exceed 30 m in height, is often confused with *P. canariensis*, which is shorter and more ornamental; but this is the only one of the genus that produces edible fruit. The leaves are pinnate, upturned, glaucous, with linear-acuminate segments and spiny stalks, and form a terminal cluster. The plants are dioecious; the inflorescences are enclosed in a spathe on a branched spadix, and pollination, generally carried out artificially for commercial purposes, is effected by shaking the male inflorescence in the spathe of the female. The flowers are inconspicuous. The fruit is a berry whose pericarp, fleshy and sweet, is the edible part. This palm can put out suckers at the base, but the stem is never branched.

Scientific name : *Phoenix reclinata*
Common name : Senegal palm
Family : *Arecaceae*

This palm, which can reach a height of 8 m, with a slender, supple stem, readily produces a number of basal suckers, and unless these are removed to permit growth the main stem remains low. The crown of leaves carried at the top is markedly downturned; the leaves themselves are pinnate, bright green, with stiff segments, acuminate and sharp at the tip, arranged in two rows. The central segment is bifid, the lower ones are spiny, the lowest so modified as to consist solely of long thorns. The plants are dioecious, with unisexual flowers growing in racemes on separate plants; many hybridizations have been attempted, sometimes successfully, especially on the Riviera in France and Italy, where the whole genus finds conditions congenial. The typical fruits are small and red, though on some hybrids they are black.

Scientific name : *Roystonea regia*
Common name : Royal palm
Family : *Arecaceae*

This robust palm grows up to 30 m or more, with a smooth erect trunk which sometimes has a swelling above the central part, though that feature may be missing. The leaves form an apical cluster; graceful and deflexed, they are unarmed, pinnate, and opposite, with an equal number of segments growing either side of the midrib. The stalk is a half-cylinder, unarmed, forming a sheath, and the whole leaf can reach a length of 3-6 m. The flowers, small and white are clustered in spadices with pendulous branches. The fruits are violet-black, globose or oblong, and the seeds are rather small.

Scientific name : *Sabal palmetto*
Common name : Caribbean cabbage palm
Family : *Arecaceae*

These palms can reach a height of 20-25 m. The stem is erect and robust, almost always covered with persistent leaf-bases and encrusted with the sheaths of dried leaves. The top is crowned by a large cluster of leaves, pinnatifid with long linear segments, partly deflexed; the long stalk, with its sheath, is erect in the central leaves, curved outward in the outer ones. The segments are thready at the edges, green above and grayish below and deeply indented at the center. The small, white flowers grown on long, branching and spreading spadices, shorter than the leaves; the fruits that follow are drupes, black when ripe and about 1 cm long. *S. palmetto* is the emblem of South Carolina and is depicted on its state flag.

Scientific name : *Sabal texana*
Common name : Texas palmetto
Family : *Arecaceae*

This palm is a robust-looking tree, 15 m or more tall with a trunk about 45 cm in diameter, reddish brown on the outside. It is one of the most elegant of the genus, though often the bases of the withered leaves, or even the leaves themselves, remain attached to the stem for a long time. The crown of foliage at the top is dense and rounded; the leaves have a long, reddish stalk and are fan shaped, with ribbed, palmate, deeply incised lamina as much as 1 m wide; the segments are acuminate, pendant and partly rotated, with slightly thready margins. The scented, hermaphrodite flowers grow in inflorescences which are generally longer than the leaves, and greatly branched. They are followed by globose, blackish fruits, drupes with prominent micropyle.

Scientific name : *Trachycarpus fortunei*
Common name : Chusan palm/Windmill palm
Family : *Arecaceae*

A palm with a strong stem covered with the sheaths of old leaves and with long, dark-brown fibres which make it look shaggy. The leaves, forming a great apical cluster, have a long, unarmed stalk and the circular lamina is divided into numerous sword-shaped, dark green segments. The flowers appear on a number of long, branching spadices, and may be unisexual or hermaphrodite on the same plant; they are small, with three sepals and three petals, and the fruits that follow them are globoid or kidney shaped, blue or bluish, with a deep furrow.

Scientific name : *Washingtonia filifera*
Common name : Petticoat palm/California fan palm
Family : *Arecaceae*

These palms, which may grow to 10-12 m have erect, cylindrical stems, wider at the base and partly covered with withered leaves which, when they fall, leave a scar on the bark, which is consequently striped in rings. The leaves have long stalks with spines almost halfway up; they are erect at first, but later they bend outward. The linear segments are spread almost like a fan and their margins have a certain number of fibrous threads. The inflorescences are ramified spadices bearing spikes, with thin, supple branches; the bracts are membraneous and glabrous. The little, white, insignificant flowers are followed by fruits that consist of a black drupe with a brown seed, rather flattened.

Broad leaved trees

Scientific name : *Acacia horrida*

Family : *Fabaceae*

The specific name denotes the viciousness of the thorns. Found in arid regions of northern and central Africa. The trunk can reach 4-7 m and has sparse, thorny branches that form a fairly regular rounded crown.

Leaves: The alternate leaves are composed of 3-7 pinnules, formed in their turn from ten pairs of linear leaflets with obtuse points, pale green, somewhat glaucescent. The stipules, transformed into spines, branch in two; they are robust, white and sharply pointed, 3-6 cm long. The scented, yellow flowers are arranged in fairly large globose, rather feathery heads growing from the leaf axis; the tree flowers in July and August.

Fruits: The fruits are big, oblong, blackish legumes containing 3-5 seeds. In its original habitat this species shows how well it can adapt to a hot, dry climate. It is similar in general appearance to *A. cavenia* but can easily be distinguished from it by the conspicuous long white thorns, especially in young specimens. In some regions of very low rainfall, in southern Italy for example, it is cultivated for its abundant summer flowers.

Propagation : By seed and by layering

Scientific name : *Acer negundo*
Comon name : Negundo, Box elder
Family : *Aceraceae*

From the Sanskrit and Bengali *nurgundi*, used for a tree that has leaves like those of the Negundo. The type species is native to the whole of North America. A tree growing upto 15-20 m high, with smooth, green young branches.

Leaves: The leaves are composed of 3-5 ovate-acuminate leaflets, deeply indented, often with the terminal leaflet trilobal.

Fruits : The fruit as in the entire genus, consists of twin samaras set at an acute angle; the wing generally looks curved inward, concave on the side facing the central axis. This very hardy tree was first imported into Europe from its homeland in 1688. Varieties of it are cultivated for the beauty of the crown in parks and gardens; *A. variegatum* has leaves with a broad white border that make its foliage very ornamental. The trees can be used as windbreaks, and a sweet liquid can sometimes be obtained by making an incision in the trunk, as with its relative, *A. saccharum*.

Propagation : By seed; but the variegated varieties may also be reproduced by grafting.

Scientific name : *Acer palmatum*
Common name : Japanese maple
Family : *Aceraceae*

The specific name refers to the shape of the leaves, like the palm of a hand with the fingers spread. Originally from China, Korea and Japan. A tree 5-8m tall with a dome crown and glabrous twigs, stalks and peduncles.

Leaves: The leaves are divided into 5-9 lobes, oblong-acuminate, doubly serrated or incised, and always with smooth stalks. In the spring it displays a foliage delicately shaded from green to red, while in autumn it glows with more vivid colours. The flowers are erect, purplish pink, growing in corymbs.

Fruits: The fruits are disamaras with the wings set at an obtuse angle. There are many varieties, and nurserymen have tried to unite the two variable features, the colour and shape of the leaves. Very similar is *A. japonicum*, the true Japanese Maple (full Moon Maple), which differs from *A. palmatum* in having its stalks covered with a light down. No one who wants to couple elegance of shape with brilliance of colour, which may vary with the seasons, should fail to plant a Japanese Maple.

Propagation : By seed, but to retain the characteristics of the leaves it is best to use cuttings.

Scientific name : *Adansonia digitata*
Common name : Baobab
Family : *Bombacaceae*

This distinctively African tree does not grow high, at most 12 m, but its trunk is probably the thickest of any in existence, as much as 10 m in diameter.

Leaves: The palmate leaves have 3 leaflets in the juvenile stage and 5-7 when adult; the widely spaced branches form a sparse, spreading crown. The wood is spongy and saturated with water, which enables the tree to withstand drought. The trunk grows without forming annual rings. The thick gray bark contains a strong fiber used in making paper and ropes. The flowers are white, large, pendulous and solitary, carried on long stalks; after a short time the petals curve outward, reflexing to reveal the purple anthers of the many stamens.

Fruits: The fruits oblong and woody, may be as much as 30 cm long and weigh several pounds. Its acid flesh is eaten by the Africans and produces a drink like lemonade.

Propagation : By seed.

Scientific name : *Alnus glutinosa*
Common name : Common alder/European alder
Family : *Betulaceae*

The genus has retained its original Latin name. *Glutinosa* refers to the sticky feel of the young leaves. A tree up to 20-30 m high, with an erect trunk, sparse, spreading ramification forming a broad, conical crown, and rough, brown bark that cracks into slender, vertical scales in the summer.

Leaves: The simple leaves, obovate and irregularly toothed at the margins, are dark green above and lighter below. The unisexual flowers grow in catkins; the males, cylindrical and pendulous, appear before the leaves, while the females are ovoid and erect. After fertilization the female inflorescence is transformed into a stalked infructenscence in the form of a little strobile, from which small winged achenes appear when ripe. The wood, homogeneous and opaque, grows harder in water and becomes remarkably strong when permanently submerged, so that it is used for piles and hydraulic works. Its charcoal is used in the manufacture of fireworks.

Propagation : By seeds, or by cuttings.

Scientific name : *Aralia elata*

Common name : Aralia/Japanese angelica tree/Devil's walking stick

Family : *Araliaceae*

This tree, which can grow as high as 15 m, has a slightly spiny trunk, branching from the base and often putting out a number of basal suckers, by means of which it can be propagated.

Leaves: The leaves are compound, more than 1 m long, generally free of spines, with toothed ovato-acuminate leaflets, dark green, generally downy on the underside and almost sessile. The flowers are small and white, growing in umbels or panicles; the tiny fruits that follow them are drupes resembling little black berries. The tree grows very quickly. There are several varieties, among them some with smooth leaves, some with spiny leaves and a rather elegant variegate with the leaves edged in white.

Propagation : By seed, and by radical suckers and root ruttings.

Scientific name : *Betula papyrifera*
Common name : White birch/Paper birch/Canoe birch
Family : *Betulaceae*

A tree up to 20-30 m high with brilliant white bark that peels off in big strips. The American Indians used to gather it carefully and stretch it on the frames of their canoes. The young branches are downy at first.

Leaves: The oval-acuminate leaves are coarsely double toothed, downy along the veins on the lower side and borne on stalks that are also hairy. They turn yellow in autumn. The unisexual flowers have the male elements growing closely in cylindrical, pendulous catkins, while the females, also in catkins, are smaller and erect.

Fruits: The fruits grow in a cylindrical infructescence which disintegrates when ripe to allow the escape of little samaras with a membraneous wing. The oil extracted from the bark contains methyl salicylate, used to treat rheumatism and inflammations.

Propagation : By seed

Scientific name : *Brachychiton acerifolium*

Family : *Sterculiaceae*

The name comes from the Greek *brachys*, short and *chiton*, tunic, from the short, overlapping scales. A tree of remarkable size which in its homelands can reach a height of more than 30 m. It is grown particularly for the unusual shape of its leaves and showy red flowers.

Leaves: The long-stalked leaves are palmato-lobate, and each lobe is in turn lobed so as to form a kind of fan; there may be 5-7 segments, and the whole leaf can measure as much as 30 cm in diameter. The flowers grow in racemes; they have no petals but are equipped with a very sticky red calyx about 2 cm long, bell shaped, with big, pointed, open, outward-curving lobes at the center of which the staminal column appears with about ten stamens, joined to the pistils.

Fruits: The fruits are big, glabrous and dehiscent follicles on a long stalk.

Propagation : By seed or by softwood cuttings.

Scientific name : *Cocculus laurifolius*
Common name : Jamtiki bel/ Marpinki
Family : *Menispermaceae*

Native to east Asia, in the Himalayan region. An evergreen tree that can grow to 5-8 m, with a spreading, rounded, very dense crown.

Leaves: The leaves are glossy, green, oblong - lanceolate, pointed at the tip and narrower at the base, with three conspicuous veins that are distinctive. The leaves are alternate and entire, with a short stalk, and join the branch very close together. The yellowish flowers are small and inconspicuous, growing in racemes; the plants are dioecious, so that they can only fruit if specimens of each sex are present.

Fruits: The fruit are drupes. The bark contains an alkaloid called coclaurina. This plant is very tolerant of pruning, even of topiary; it can also be grown as a bush or big hedge.

Propagation : By seed, cuttings or layering.

Scientific name : *Cotinus coggyria*
Common name : Smoke tree/Venetian sumac
Family : *Anacardiaceae*

This tree, not more than 5 m high, has entire, simple leaves, rarely more than 7 cm long, oval, glabrous above and below and turning dark red, almost violet tinged, in autumn. The flowers, hermaphrodite or unisexual, have yellowish-green petals which fall early and grow in a panicle with threadlike, branched, downy peduncles.

Fruits: The fruits, small oval drupes, red-brown and shiny since they are glabrous, are beautiful but poisonous; in autumn they remain hanging on the hairy stalks, giving the whole tree a feathery look which has earned it the name of Periwig Tree. It is cultivated as an ornamental in gardens.

Leaves: The leaves of the tree provide a tannin used in the tanning of high quality leather. The white sapwood and yellowish-red heartwood gives a fine polish but can be used only for small objects.

Propagation : By seed, suckers, or cuttings.

Scientific name : *Cunonia capensis*
Family : *Cunoniaceae*

The name commemorates John Christian Cuno, a Dutch naturalist of the second half of the 18th century. Native to South Africa, the tree grows 6-10 m tall.

Leaves: It has evergreen leaves, pinnate and leathery, with two or three pairs of oblong, lanceolate leaves, with glossy surfaces, toothed margins and reddish stalks. The white flowers are borne in dense axillary racemes which look like spikes; they are small, with a short calyx tube, five petals and ten very prominent stamens.

Fruits: The fruit is a leathery capsule. Since the tree branches from the base up it can also be cultivated as a shrub, and small specimens can be grown as house plants if the temperature is not too high.

Propagation : By softwood cuttings.

Scientific name : *Diospyros ebenum*

Common name : Ebony tree

Family : *Ebenaceae*

Derived from the Greek *diospyros*, a word compounded from *dios*, divine, and *pyros*, grain, referring to the edible fruit of some species; the specific name is from the Greek *ebenos*. Occurs in Sri Lanka, Malaysia, India and Indo-China. The tree that grows very tall, especially in Sri Lanka, though of smaller size in Indo-China.

Leaves: It has gray bark, and the alternate, very short-stalked leaves are ovate-oblong, leathery, entire and glossy. The flowers are sessile, axillary and solitary.

Fruits: The fruit is an oval berry. The wood is black and heavy, so hard that it is almost impossible to drive a nail into it, of very fine grain and able to take a perfect polish. A valuable wood, it is reserved for luxury goods and veneers; it is also used for the manufacture of musical instruments such as flutes and clarinets and pianoforte keys. Also included under the commercial term "ebony" are many other species of *Diospyros*; and the wood is widely imitated by dyeing black common fine-grained woods, such as pear.

Propagation : By seed, which must be planted deeply.

Scientific name : *Disanthus cercidifolius*
Common name : Witch- hazel
Family : *Hamamelidaceae*

From the Greek *dis*, twice, and *anthos*, flower, referring to the paired flowers. A small deciduous tree that grows to 3 m, often polycormic; the only species of its genus.

Leaves: The alternate leaves on long stalks are entire, almost heart shaped, blue green on the upper surface and lighter beneath, some 10 cm long; before they drop in autumn they take on colours ranging from red to orange. Despite its flower, which grows in pairs united at the ball and set opposite on erect stalks, the plant is grown essentially for its attractive foliage. The small flowers are dark purple, with a calyx divided into five segments and five slender, open petals; the five stamens are shorter than the sepals.

Fruits: The fruit is a dehiscent capsule with several shiny black seeds in each loculus.

Propagation : By seed; germination takes two or three years. Alternatively, by grafting on *Hamamelis* seedlings.

Scientific name : *Eucalyptus camaldulensis*
Common name : Red gum
Family : *Myrtaceae*

The name comes from the Greek *eu,* well, and *kalyptos*, covered: the mouths of the flowers are covered by a woody operculum. This big tree, called Red Gum from the colour of the resinous sap (the name Blue Gum being applied to *E. globules*), can grow more than 60 m high in its country of origin, though it more often stops at 30-40 m. The bark of the adult trunk is gray, and is shed and replaced; if persistent, it flakes off in thin strips. Bark of young trees is reddish.

Leaves: The young leaves are ovate, but long leaves appear later, lanceolate and acuminate, which may grow as much as 30 cm long. The woody calyx is hemispherical, while the operculum the "lid" which closes the flowers, is conical; when it opens and falls, white stamens emerge. The reddish wood is disease resistant though not very strong, and hard to work when dry. Bees are much attracted by the flowers.

Propagation : By seed.

Scientific name : *Ficus lyrata*
Common name : Benjo fig/Fiddle-leaf fig
Family : *Moraceae*

The specific name refers to the shape of the leaves, which suggest a stringed instrument. Native to tropical West Africa. Young specimens of this plant are often used as ornamental house plants, but in its natural surroundings the tree can reach a height of 12 m, with an erect trunk often divided from the base, and not very spreading branches, with more erect habit.

Leaves: The glossy, leathery leaves are very big, sometimes as much as 50 cm long, and have a curious shape that recalls a stringed instrument, with an eared base above which the leaf blade first narrows and then broadens again; the margin is obtuse at the top, with a short, cuspidate point at the tip of the midrib; the margins are sinuate. The inflorescences, typical of the genus, are pyramidal, with unisexual flowers enclosed in hollow receptacles. They are followed by fruit of the syconial type, with a green skin with white dots.

Propagation: By layering.

Scientific name : *Ficus sycomorus*
Common name : Sycamore
Family : *Moraceae*

In Greek, *syka* means fig. A tree of erect, spreading habit, with a greatly branched, rounded crown, up to 10-13 m in height; the trunk is a fulvous yellow.

Leaves: The leaves, partially deciduous, fall for some months every year, are ovate, with a rough surface and downy veins, glaucescent when young, leathery and olive green when mature. The flowers are greenish. The small, edible fruits are produced in great quantities and, as in all the genus, are syconial. The wood is valuable for its hardness, so much so that in ancient Egypt it was used to make sarcophagi.

Propagation : By cuttings or layering.

Scientific name : *Fraxinus excelsior*
Common name : European ash
Family : *Oleaceae*

The Latin specific name means, literally, "higher", compared presumably with *F. ornus*, which is common in the woods of Italy and may also have been known to the Romans. Widely distributed in the Northern Hemisphere, the ash grows wild in much of Europe at Altitudes of up to 1500 m (5000 ft), rarely forming monocultural forests. This tree grows up to 10 m high, terminating in a sparse and irregular crown. The red-brown bark cracks with age.

Leaves: The compound leaves are formed from 9-15 oval-lanceolate leaflets with toothed margin. The flowers are not very conspicuous, having no calyx or corolla; they grow in panicles that are set erect at first but later become pendulous, and open before the leaves appear to facilitate pollination, which is anemophilous.

Fruits: The fruits are samaras, wind scattered by their single wing.

Propagation : By seed.

Scientific name : *Gleditsia triacanthos*
Common name : Honey locust/Sweet locust
Family : *Fabaceae*

A deciduous tree which grows more than 30 m high, with a broad crown.

Leaves: The leaves, which appear rather late, are alternate and may be pinnate with 10-12 pairs of leaflets, or bipinnate with 4-7 pairs of leaflets; both forms appear on the same plant and are pubescent along the midrib. The leaflets are oblong-lanceolate, with serrated margin, and turn yellow before they fall. The leaves fall early, so that the foliage is only on the tree for a short time. Long, vicious thorns guard the trunk and branches. The insignificant flowers grow in greeny-yellow panicles, with a small calyx and 3-5 petals.

Fruits: The fruits are brown, indehiscent pods, curved and twisted, up to 45 cm long, with depressions between the seeds. They grow very profusely in clusters. It is an excellent shade tree for terraces.

Propagation : By seeds, which are rather hard and must be soaked in water before sowing.

Scientific name : *Grevillea robusta*
Common name : Silk-Oak Grevillea
Family : *Proteaceae*

The name commemorates Charles F. Greville (1749-1809), one of the founders of the Royal Horticultural Society of London. In its native climates this tree reaches a height of some 40 m although, particularly in the North, it is often cultivated in the very young stage as a pot plant.

Leaves: The leaves are bipinnatifid, like big fern fronds, green on the upper side, silvery below. The flowers grow in racemes on the old wood of the previous year's branches; they are apetalous, with a short calyx of four upturned lobes, four stamens with sessile anthers and a long, curved style. They are orange and rather small, but they grow densely on the branches of the inflorescence to form striking clusters.

Fruits: The fruits are follicles, with 1-2 flat winged seeds. The timber is tough and elastic, used in joinery. The flowers are very attractive to bees making this an important honey plant.

Propagation : By seed or by cuttings.

Scientific name : *Gymnocladus dioicus*
Common name : Kentucky coffee tree
Family : *Fabaceae*

The name comes from the Greek *gymnos*, bare, and *klados*, branch, referring to the large branches which remain bare from early to late winter.

Native to the United States, in a central eastern strip northwest of the Allegheny Mountains, this tree reaches a height of over 30 m, and has a stout trunk with a great number of erect branches and a globose crown.

Leaves: The leaves are as much as 90 cm long and 45 cm wide; alternate, rather irregularly bipinnate, with 3-7 pairs of leaflets; they are oval or ovate and acuminate, and downy in the juvenile stage. The stalks persist after the leaves have fallen. The small flowers grow in greenish-white inflorescences and may be unisexual or polygamous.

Fruits: The fruits are legumes, very large and conspicuous, which persist during the winter; the seeds were once used as a substitute for coffee and hence the common name of Kentucky Coffee Tree.

Propagation : By seed or by radical shoots.

Scientific name : *Halesia monticola*
Common name : Mountain silverbell
Family : *Styracaceae*

Grows wild in the mountains of the south-eastern United States. This tree reaches some 25 m, with a pyramidal crown, and grows rapidly.

Leaves: The deciduous leaves are entire, stalked but without stipules, oblong-ovate and toothed at the margins, and glabrous; they turn yellow before they fall. The flowers have united petals and grow in axillary clusters in spring on the previous year's wood; they have a small calyx and a fairly big, bellshaped white corolla with four lobes and are generally pendulous.

Fruits: The fruits that follow them are drupes with four big longitudinal wings; the stone contains 1-3 seeds.

Propagation : By seed, by softwood cuttings and by layering.

Scientific name : *Ilex aquifolium*
Common name : English holly
Family : *Aquifoliaceae*

Native to Britain and also grows in central-southern Europe, from the Iberian Peninsula to the Caucasus and as far as Persia and North Africa. A tree 8-10 m high, with a smooth bark, green in young plants, then gray. The crown is a pyramidal cone.

Leaves: The leaves are persistent, oval or elliptical and pointed, with spines at the margins and the point; they are leathery, alternate, glossy dark green above but lighter and duller on the underside. The unisexual flowers are small and white, with short stalks.

Fruits: The fruit is a bright-red globose or ovoid drupe which contains four seeds and has purgative properties. The wood is hard and can be dyed; it may be used as a substitute for ebony. The bark is sticky and contains tannin.

Propagation : By seed or by cuttings of partially lignified shoots.

Scientific name : *Juglans nigra*
Common name : Eastern black walnut
Family : *Juglandaceae*

A deciduous tree from 20 to 40 m in height, with an erect trunk with very dark, deeply furrowed bark and rounded crown.

Leaves: The pinnate leaves are about 60 cm long and have 10-20 lanceolate-acuminate leaflets, glabrous on the upper side and pubescent on the under side. The flowers are unisexual, both sexes being carried on the same plant; the females grow in small, sparsely flowering racemes, and both sexes are small and insignificant.

Fruits: The fruit is a drupe in which the seed, enclosed in a woody stone, is surrounded by a fleshy husk. The fruits may be either solitary or paired; they are very much more spherical in shape than those of *J. regia*, and the nut is small and very woody. The tree is cultivated not only for ornament but also for its very beautiful dark red wood, hard and proof against attack by insects. The roots contain a substance called juglone which can poison surrounding trees.

Propagation : By seed.

Scientific name : *Kigelia pinnata*
Common name : Sausage tree
Family : *Bignoniaceae*

A tree with a stout trunk, often branching from the base, which grows 6-15 m high, with spreading branches and pendulous secondary branchlets.

Leaves: The leaves are imparipinnate, with 7-9 oblong leaflets, entire or serrate, smooth on the upper side, often slightly downy below. The lateral leaflets are sessile on the midrib while the apial leaflet grows out as much as 30 cm from the final pair; as a result, the crown of the tree does not look very dense. Pendulous racemes bear large, showy flowers with a tubular corolla, yellow at the base and red at the open lobes. The main branch of the inflorescence is very long, and the fruit that grows on it, brownish-gray and cylindrical, hangs from a stalk that may be as much as 1 m long. Some tribes make a drink from it.

Propagation : By seed.

Scientific name : *Laurus nobilis*
Common name : Laurel/ Sweet bay
Family : *Lauraceae*

The name is the same as the Latin. The Graaks called the plant *daphne* after the legend that the nymph Daphne was turned into a laurel to escape from Apollo. The tree became sacred to that god and was used in festivals in his honor, and when the cult passed to Rome, conquerors and poets were crowned with laurel, a custom that survived for a very long time. Even Napoleon's crown featured a golden laurel wreath. Grows wild in the Mediterranean region. This evergreen tree has a maximum height of 8-12 m and bushy habit.

Leaves: The lanceolate leaves are glossy green, alternate, leathery and aromatic. The flowers are unisexual, yellowish, growing in small umbels. The plants are dioecious flowering in spring, and the little flowers on the female plants are followed by small black drupes, glossy and ovoid or globose, rather like olives. There are several varieties, including *crispa* and *angustifoli*.

Propagation: By seed, cuttings, layering and basal suckers.

Scientific name : *Morus alba*
Common name : White mulberry
Family : *Moraceae*

A tree 10-18 m high, with an irregularly branching trunk forming a spreading crown. The bark, gray at first, later becomes brown and cracks longitudinally.

Leaves: The leaves deciduous and alternate, have a channeled stalk with deciduous stipules; the leaf blade, ovato-acute and three lobed, is smooth and light green on both sides. The flowers are unisexual, the males growing in cyclindrical catkins with a short stalk, the females in subglobose catkins.

Fruits: The fruit, white, pink or violet, resembles the Black Mulberry's and is sweet even when unripe. Introduced into the Mediterranean probably during the 12th century together with the silkworm, which lives exclusively on its leaves, the tree was a characteristic feature of the Italian landscape when silkworm breeding was both an industry and a home craft. There is now a variety *pendula* which is grown in gardens.

Propagation : By seed or cuttings, but mostly by grafting.

Scientific name : *Morus nigra*
Common name : Black mulberry
Family : *Moraceae*

The specific name refers to the dark colour of the fruit compared with that of the related *M. alba.* Native to Asia Minor and Persia, this tree too was introduced into the Mediterranean region many years ago. Generally similar to the White Mulberry, but more robust, with thicker, rougher trunk and branches, the tree can reach 10-15 m and has a denser crown.

Leaves: The leaves with a shorter, not deeply channeled stalk, are markedly cordate at the base, undivided or with 3-5 lobes, stiffer and of a darker green on the upper side, while the underside is downy. The flowers are unisexual.

Fruits: The fruits are larger than those of the White Mulberry, composed of a number of little violet-black drupes and unlike the White Mulberry, sour until fully ripe. The tree is grown for its fruit, which is used to make slightly tart jams and jellies. The foliage is rough and not much used for feeding silkworms.

Propagation: By seed, cuttings, layering or grafting.

Scientific name : *Nolina longifolia*
Family : *Agavaceae*

A woody, slow-growing plant which becomes arborescent with time and can grow as much as 3 m tall. The thick trunk, covered in a very thick corky bark, has rather few erect branches, each with a cluster of long leaves at the tip, linear and acuminate, with saw-toothed margins. The young branches in the centre are erect, but on maturity they become extroflex and later completely pendulous, remaining persistent for a long time even after they are withered. The creamy white bell-shaped flowers are very small, growing in long, dense, erect panicles from the spreading ramification; they may be unisexual and polygamous.

Fruits: The fruit contains one to three seeds.

Propagation: By seed, which is very slow-growing; if suckers appear they are a preferable means of reproduction.

Scientific name : *Phellodendron amurense*
Common name : Amur cork tree
Family : *Rutaceae*

Derived from the Greek *phellos*, cork and *dendron*, tree, since the bark is corky.

A tree 12-15 m tall, with thick, spreading branches and a distinctive form. The bark, deeply fissured, thick and corky, is light gray at first but becomes darker.

Leaves: The leaves are more than 30-50 cm long, compound, imparipinnate and alternate, composed of 5-11 oval leaflets with long points at the tip, and ciliate. Seen against the light they show transparent dots due to oleriferous cells which contain aromatic elements; when rubbed between the fingers they smell of turpentine. Light green in autumn, they turn golden yellow before they fall. The small, whitish flowers; insignificant and unisexual, open in late spring and grow in corymbs.

Fruits: The fruits are black drupes the size of a pea, ripening in the autumn; they too smell of the turpentine when rubbed. The wood, with white sapwood and brown heartwood, is used in China for making furniture, gunstocks and small tools.

Propagation: By seed or cuttings taken in July.

Scientific name : *Populus alba*
Common name : White poplar
Family : *Salinaceae*

Under ideal conditions this tree can reach a height of 30 m. It has a gray to gray-green bark on a generally erect trunk; when young it has bands of lenticels, later it becomes fissured and rough.

Leaves: The adult leaves have short stalk with ovate or elliptical leaf blade, sinuate or bluntly denate at the margin, with a gray tomentum on the underside and intense green above; on young shoots the leaves are roughly triangular or palmate-lobate, with a long stalk. The flowers are lateral catkins, the males cylindrical with the stamens and anthers purple at first turning to yellow.

Fruits: The fruit is a glabrous capsule. The wood is mediocre, only used for planking and packing cases or as supporting parts in furniture manufactures from finer woods.

Propagation : By seed, radical suckers or cuttings.

Scientific name : *Quercus robur*
Common name : Common oak/English oak
Family : *Fagaceae*

The specific name is a Latin word used to describe all hard wood, especially that of the oak. A majestic tree which can reach a height of 50 m, with a stout trunk and large, irregular crown.

Leaves: The alternate, simple, deciduous leaves, herbaceous rather than leathery, are ovate-oblong in the adult state, narrower at the base and broader towards the tip, with 5-7 pairs of broad lobes separated by roundish bays; the colour is dark green and glossy on the upper side, paler below. The acorn is ovate-oblong and pointed, partly enclosed by a cupule of appressed scales, overlapping and slightly tomentose. The wood is favorite of shipbuilders and for use as roof timbers, the so-called Slavonian Oak used for making casks for the aging of valuable wines and brandy, comes from the Common Oak and not from the Sessile Oak, which grows in Slavonia. The wood is easy to work and long lasting.

Propagation: By seed.

Scientific name : *Salix alba*
Common name : White willow
Family : *Salicaceae*

A tree up to 20-25 m tall, with a maximum diameter of about 60 m. The crown is spreading, composed of long, rising branches, the old ones silvery gray, the young branchlets long and flexible with a reddish-green or brown bark. The bark of the trunk is gray and cracks with age.

Leaves: The leaves with narrow stalks lacking glands and bearing deciduous stipules, have lanceolate, acuminate blades 7-10 cm long, narrowing toward the tip with the greatest width in the lower half, finely toothed margins and glandular; they are green and slightly glossy on the surface, lighter below, and their appressed hairs make them look silvery and silky. Flowers appear at the same time as the leaves, the males up to 7 cm long with two stamens to the yellow anthers, the females thinner and lightly stalked. Flowering occurs in March-April.

Fruits: The fruits are subsessile, glabrous capsules; the seeds have a white pappus.

Propagation: By seed; also by cuttings.

Scientific name : *Salix babylonica*
Common name : Weeping willow
Family : *Salicacea*

The specific name suggests that the tree may have come from Babylon, possibly a way station between Europe and the tree's true home is central Asia. Other authorities connect the name with the Jews' bitter complaint during the Babylonian captivity; 'We hanged our harps upon the willows in the midst thereof'. A tree growing to 8-10 m with a distinctively shaped crown due to the presence of long, thin, pendulous young branches which may droop down to the ground.

Leaves: The leaves, borne on a short stalk, are narrowly lanceolate, with a long terminal point, serrated at the margin and very often glaucescent on the underside, male and female flowers. Widely grown in parks and gardens, especially near lakes, where the tree looks most romantic when reflected in the water. Rather rare as a true species; those commonly seen are most often hybrids.

Propagation: only by cuttings of female trees.

Scientific name : *Sambucus nigra*
Common name : European Elder
Family : *Caprifoliaceae*

The name has kept the original Latin form, probably derived from *sambuca,* a musical instrument made from elder wood. The species is widely distributed in Europe and Asia, and may grow in the Alps at altitudes up to 1500m. The elder is a small tree, growing to no more than 3-5 m, although exceptional specimens have been met as tall as 12 m and 40 cm in diameter. The bark of both trunk and branches is light gray, with warty brown lenticels; with age it cracks and takes on a horny, pimply look.

Leaves: The deciduous, opposite leaves are pinnatisect, composed of 3-7 ovate-acuminate, toothed leaflets, light green and downy; the stalk is thick, short and widening at the base. The strongly scented hermaphrodite flowers grow in large terminal inflorescences with a flat surface.

Fruits: The fruits are violet-black drupes with red juice, containing three oval brown seeds. The branches contain a great deal of white pith which is used in microscopy to enclose small sections, enabling them to be sliced more easily to the desired thickness of a few microns.

Propagation : By seed or by hardwood cuttings.

Scientific name : *Sophora japonica*
Common name : Pagoda Tree
Family : *Fabaceae*

A deciduous tree that can reach a height of over 20 cm and whose trunk, as it grows old, often becomes curiously twisted and knotted. The first specimen planted in Europe, in Kew Gardens, in the second half of the 18th century, has actually assumed a semiprostrate posture. The branches are wide and the crown looks rounded.

Leaves: The leaves are compound, imparipinnate, with 7-9 stalked lanceolate acuminate leaflets, dark green and slightly downy on the underside. The flowers, creamy white, appear in long panicles; the individual flowers are 1 cm long and papilionate, rather like those of the pea; the fruits are little berries strung together like a necklace. Flowers and fruit contain a yellow colouring matter. The variety *pendula*, of weeping habit, is particularly ornamental.

Propagation : By seed, the varieties are grafted on seed stock by cleft grafting.

Scientific name : *Sophora secundiflora*

Family : *Leguminosae*

The species name comes from the Latin 'secundiflorus', from the flowers which are turned to one side. The tree comes from the southern United States, from Texas to New Mexico. An evergreen growing up to 10 m high, with a slender trunk and erect ramification that tends to make it look tall.

Leaves: The compound leaves, 15 cm long, have 7-9 rather slender leaflets, elliptical and rounded or emarginated at the top, covered with a silky down when young, almost sessile, with a wedge-shaped base. The violet-blue, strongly scented flowers are papilionaceous, more than 2 cm long; they grow in terminal racemes about 10 cm long and are all turned to one side. They appear in spring, unlike those of the species *S.japonica*, which flowers much later, at the end of the summer.

Fruits: The fruits are moniliform, woody pods with a whitish tomentum; the scarlet seeds contain a poisonous alkaloid called sophorine.

Propagation: By seed.

Scientific name : *Tamarix pentandra*

Common name : Tamarisk

Family : *Tamaricaceae*

The specific name refers to the five very conspicuous stamens, the red end of the anthers standing out against the rosy corolla. Native to Mediterranean Europe, though it is sometimes confused with species from China and Japan. This tree can reach a maximum height of 5 m; the trunk and branches have lavender-tinged bark.

Leaves: The blue-green alternate leaves are sessile, small and scalelike. The inconspicuous pink flowers grow in racemes that appear at the beginning of the summer and form plumes at the tips of the branches.

Propagation: Propagation by seed is carried out in spring or autumn but is not effective.

Scientific name : *Tilia americana*

Common name : Basswood/ American Linden

Family : *Tiliaceae*

The American Linden is a beautiful tree which in its natural range reaches a height of 40 m or more.

Leaves: The leaves may be as much as 22-23 cm long; simple alternate and broadly ovate; toothed at the margin and the base asymmetrically cordate, terminating in a sharp point. The yellowish-white flowers appear at the beginning of summer, clustered at the end of a stalk which also bears a wide, membraneous bract important in dissemination. They have five sepals and five petals, also five staminoides, not found in any of the other species. The wood is soft, white and easy to work. The American Linden is grown as an ornamental, especially for the shape of its slightly trilobate leaves. It is an important honey plant.

Propagation : By seed, by layering or by cuttings.

Scientific name : *Ulmus campestris*
Common name : English Elm
Family : *Ulmaceae*

Leaves: This tree, 30 m tall, has alternate simple leaves with long points at the tip, asymmetrical at the base, with a toothed margin. The flowers appear before the leaves and have red-purple anthers.

Fruits: The fruit, which also appears early compared with the leaves, is a broad-winged samara well adapted for dissemination. The bark is smooth and brown-gray when young, later becoming cracked. It is fibrous, and gardeners use it to make ties for grafts. The leaves make excellent cattle food on account of their high protein content; for this reason the English Elm is preferred to other species where grass is sparse and cattle are grazed off trees. It is also widely grown as a roadside ornamental. The elm has a prolonged rest period, which protects the woody buds from late frosts even though, as a result of early flowering, the production of seeds is sometimes adversely affected by long cold periods.

Propagtaion : By seed, and by suckers.

Scientific name : *Zelkova crenata*
Common name : Zelkova
Family : *Ulmaceae*

This is a beautiful tree, up to 30 m tall, with a thick trunk that divides into numerous slender vertical branches. The smooth gray bark comes off in little scales rather like that of the plane.

Leaves: The small, alternate leaves have short stalks, a toothed margin and downy veins on the underside. The insignificant green flowers are unisexual but are all borne on the same plant, they open in April at the same time as the leaves and are strongly scented.

Fruits: The fruit is a small nut about the size of a pea. A slow growing tree, it produces a hard, compact, resilient orange-yellow or golden-brown wood, used for high-quality furniture and in ships and railroad cars, as well as for gunstocks, introduced into Europe in 1760, it can be distinguished from the elm, which it somewhat resembles by its smooth bark.

Propagation: By seed or by grafting onto elm stock.

9

Trees for Bonsai

Bonsai first appeared in China over a thousand years ago on a very basic scale, known as *pun-sai,* where it was the practice of growing single specimen trees in pots. These early specimens displayed sparse foliage and rugged gnarled trunks which often looked like animals, dragons and birds.

With Japan's adoption of many cultural trademarks of China, bonsai was also introduced to Japan during the Kamakura period (1185-1333) by means of Zen Buddhism, which at that time was rapidly spreading around Asia. Once bonsai was introduced into Japan, the art was refined to an extent not yet approached in China. Over time, bonsai, which was confined to the Buddhist monks and their monasteries, were introduced as a symbol of prestige and honour. It became an important part of Japanese life by being displayed on specially designed shelves. For the Japanese, bonsai represents a fusion of strong ancient beliefs with the Eastern philosophies of the harmony between man, the soul and nature.

Over time, bonsai began to take on different styles, each of which varied immensely from one another. Other culturally important elements were in the bonsai plantings such as rocks, supplementary and accent plants, and even small buildings and people which itself is known as the art of *bon-kei.* They also looked at reproducing miniature landscape in nature known as *sai-kei* which further investigated the diverse range of artistic possibilities for bonsai. In the mid 19th century, the art of bonsai was introduced to the Western world. Bonsai has now evolved to reflect changing tastes and times with a great variety of countries, cultures and conditions in which it is now practiced.

The art of Bonsai

A bonsai is not a genetically dwarfed plant and is not kept small by cruelty in any way. In fact, given an adequate supply of water, air, light and nutrients, a properly maintained bonsai should outlive a full size tree of the same species. Although bonsai are small in comparison to their huge life-sized brothers, most are over 25 centimeters tall and up to 1 metre in height. Bonsai can be developed from seeds or cuttings, from young trees or from naturally occurring stunted trees transplanted into containers. Bonsai are kept small and trained by pruning branches and roots, by periodic repotting, by pinching off new growth, and by wiring the branches and trunk so that they grow into the desired shape. Japanese tradition holds that three basic virtues are necessary to create a bonsai-*shin-zen-bi,* standing for truth, goodness and beauty. Given proper care, bonsai can live for hundreds of years, with prized specimens being passed from generation to generation, admired for their age, and revered as a reminder of those who have cared for them over the centuries

Bonsai are ordinary trees or plants. Small leaved varieties are most suitable, but essentially any plant can be used, regardless of the size it grows to in the wild. In Japan, varieties of pine, azalea, camellia, bamboo and plum are most often used. The artist never duplicates nature but rather expresses a personal aesthetic philosophy by manipulating it. The bonsai may suggest many things, but in all cases must look natural and never show the intervention of human hands.

Containers for Bonsai

When it comes to bonsai, it is not just the plant and its styling that makes the bonsai appealing to the viewer. The choice of the container is quite important to the overall look and 'feel' of the tree. The purpose of the container is to provide suitable accommodation for the tree's needs and to compliment its branch structure and seasonal or year round colours. Practical and aesthetic factors affect the choice of pot.

The pot must hold enough soil for the roots to develop over a year or two. It should be frost-proof with enough drainage holes. Bonsai containers are usually shallow, but sometimes you may use a deeper one to hold a fruiting tree, for example, that needs plenty of water to swell its fruits. It is said that unglazed, dark-coloured containers are usually chosen for classic bonsai or to impart a look of age. Evergreens such as conifers, look best in brown pots, but glazed containers should be used for flowering trees or trees with unusual colours or characteristics. Over the years, it has become traditional to select rather neutral shades of brown, grey and red for evergreens and the more colourful pots of green, blue and white for deciduous trees. It has also long been recognized that flowering trees look good in green, blue and deep purple pots. The exception to this is that trees and shrubs with red flowers look best in white pots. It is considered the norm for trees with red fruit in white pots, and trees with yellow or orange fruits in blue pots. Basic brown pots are always correct. Generally, the more mature and aged the bonsai, the plainer the container should be.

Keep the container in scale with the tree, an individual tree should not look lost in a large pot, nor overwhelm a tiny one. As a guideline, a mainly vertical tree needs a pot with a length between two-thirds to three quarters of the tree's height. The pot's length should be two-thirds and three-quarters of the overall width of a strongly horizontal tree. Certain styles and types of tree demand deeper pots. Pots with glazed interiors should be avoided.

Nutrition of bonsai

To keep the bonsai in good shape and to maintain healthy growth requires regular feeding at the right time of year with the correct fertilizer. Fertilizer must be continually replenished due to a certain amount being washed out each time the plant is watered.

Slow-release fertilizer which will take care of feeding requirements for several weeks at a time is probably the type of fertilizer most used for bonsai. Soluble powder and liquid feeds are also used for rapid effect over short-term periods. These are applied to the soil with a watering can and are used up relatively quickly. A balanced fertilizer with fairly low nitrogen content will be ideal.

Repotting bonsai

Regular reporting is vital to the health and growth of Bonsai. Generally, the trees should be repotted at a time when they are most dormant. Young or small bonsai requires repotting every two or three years, and older and larger specimens less often. A bonsai needs repotting if water takes a long time to drain through the soil or if the roots are crowding around the sides.

To prune the roots, use very sharp cutters. Start by cutting the thick, old brown roots that have come close to the edge of the pot and are restricting the growth of the young 'feeder roots'. Remove a third to a half of these roots. The thinner roots which hang below the depth of the pot are trimmed into a suitable shape that the pot will accommodate. To balance out the extensive pruning done on the roots the branches of the bonsai should also be pruned so that it can recover quicker.

Watering bonsai

The frequency of watering bonsai depends on many things. A dry wind, excessive heat, or a combination of both, can quickly dry out the soil. In summer the bonsai are watered every day in the summer and every three days in rainy season. A drip system is ideal. A watering can with a fine rose attached to water the soil and roots can also be used. Using a general water sprayer to increase humidity is also a good idea in dry periods.

Wiring bonsai

When wiring, try to imitate the natural curves of trees in nature. Branches those are unlikely to break when pressure applied to twist the wire around the branch are wired. There are two types of bonsai wire available, copper wire and aluminum wire. Aluminum wire which is usually coloured to look like copper wire is much more malleable than copper wire and generally has the same effect. The size of wire used depends on the size of branch to be trained and which is dependent on how significantly we want to change the shape of a branch and how stubborn the species of plant is. For pencil-thick sized branches the recommended gauge is 3.5 mm aluminum wire. The wire is removed carefully after the branch is trained to the desired shape.

Bonsai styles

The five basic bonsai styles are Formal upright, Informal upright, Slanting (or windswept), Semi-cascade and Cascade. All have their own beauty and serenity. In *Formal upright style* the tree trunk is straight. In *Informal upright style* in nature, a tree bends or alters it direction away from wind or shade of other trees or buildings, or towards light. In an informal upright bonsai the trunk should slightly bend to the right or left, but never towards the viewer. In *Slanting Style,* whether curved or straight, the whole trunk leans at a definite angle. The stronger roots grow out on the side, away from the angle of the trunk lean, to support the weight. Trees that slant naturally occur as a result of buffeting winds or deep shade during early development. In *Semi-Cascade style* the growing tip reaches below the base of a container. The trunk has a natural taper and gives the impression of the forces of nature pulling against the forces of gravity. Branches appear to be seeking the light. The winding main trunk is reminiscent of a stream meandering down the side of a mountain. In *Cascade style* the trunk which is tapered grows down below the container and gives the impression of the tree being forced down by the forces of gravity. The tree trunk usually also twists as if to emulate a meandering steam with elegant alternating branches protruding from it.

Trees suitable for bonsai

***Adenanthera pavonina* (Coral wood tree)**

A handsome tree with green bipinnate foliage. The flowers are small, creamish white and magnificient. Thrives best in humid localities. It is mostly trained into a formal upright style.

Bombax malabaricum **(Silk cotton tree)**

It is a deciduous tree with wide spreading branches. The large bright crimson or red flowers are borne in clusters on the branches. The tree is often found growing in nature through seed and young plants can be collected for future bonsai.

Brassia actiniphylla **(Umbrella tree)**

A handsome medium growing evergreen tree with large radially arranged leaves and has few branches. The spikes are handsome and bright scarlet or coral-red in colour borne terminally and pointing radially.

Butea monosperma **(Flame of the forest)**

A small to medium sized, untidy deciduous tree with crooked trunks and branches. The leaves are trifoliate, leathery and rough. The flowers are large and crowded on leafless branches. The flowers are bright flaming scarlet orange with dark olive-green to black calyces, with the shape resembling a parrot's beak. The flowering season is between April and May. The tree can withstand frost and drought very well. It is suitable for a wide range of bonsai styles - formal, informal and multi trunk. The tree is propagated by seeds.

Callistemon lanceolatus **(Red bottle brush)**

A dwarf evergreen tree with slender drooping branches and linear leaves, grayish brown, rough bark with narrow ridges add illusion in bonsai. It is best known for its bright red bottle-brush-like flowers. *Callistemon* is suitable for a wide range of bonsai styles-formal, informal, slanting, multi-trunk and raft. Potting and repotting are preferred in early rains. Wiring is done in late summer. Propagation is by seed, cutting or air-layering. Flowers can be expected from three to four years from seed-grown plants.

Casuarina equisetifolia **(Casuarina)**

An attractive tree of Australian origin which can be used freely in bonsai. The foliage is reduced to mere scales, with the branchlets having the appearance of needles of pine. It also produces cones, female cones borne at the base of branchlets. It is a hardy tree and can tolerate severe pruning. Casuarina is mostly trained into a formal upright style and wiring can be done from the beginning of rain. Best time for pruning is late summer. Repotting may be needed every two or three years. It is easily grown from seeds; semi hardwood cuttings have also been successful.

Cedrus deodara (Indian cedar, Deodar)

A large handsome tree, suitable for the hills. A hardy plant requiring plenty of sunlight. Leaves bluish green and attractive. Potting can be carried out almost throughout the year. It is a slow growing plant and can be trained into different styles.

Cinnamomum camphora (Cinnamon)

A medium-sized evergreen tree, mostly used in informal and slanting, twin trunk and clump style. Multi-trunk style can easily be induced simply by cutting at the soil level. Such pruning should be carried out in between late winter and early rains. Potting should be done in early rains, while wiring can be applied at any time of the year except in winter. Propagation is done through layering.

Citrus aurantifolia (Lime)

A shrub or small tree, having aromatic, attractive leaves. It is thus suggested to grow as a medium-sized or large bonsai, which in turn will give abundant edible fruits.

Citrus reticulata (Mandarin)

A small tree, cultivated mainly for its edible fruits. The young twigs angled, later cylindrical with spines in leaf axils, older branches often spineless, leaves unifoliate, petioles winged and articulated with leaf blade. Flowers single in leaf axils or in short racemes and are white in colour. In mandarins, the fruit peel is loose and easily detached from the segments. It can be trained into formal and informal styles.

Cordia sebestena

A native of Peru, small evergreen tree or large shrub, often with straight trunk, but sometimes branching from base, its suggested uses for bonsai are from upright group planting to clump styles. Dark grey bark, rough, marked with deep longitudinal furrows is an extra attraction of this plant. Propagation is from seed, cutting or layering. A quick-growing plant. Seedlings come to flowering in two years. Flowers appear round the year, but profusely from January to March. Severe pruning is possible in many months of the year and for repotting, the best time is post flowering.

Diospyros peregrina **(River ebony)**

An evergreen ornamental tree of moderate size with short, straight trunk. The added advantage is that the tree looks very beautiful when red leaves appear in March-April. The fruit subglobose and may be up to 5 cm or more. Many fruits on a miniature plant may be disproportionate but the imbalance can be dealt with by limiting the number of fruits. Usually three fruits are allowed and placed in such a way to form a triangle with the apex. Informal upright form is the common style, raft style is also possible. Propagation is done by seeds or layering.

Erythrina crista-galli **(Common coral-tree)**

It is a dwarf tree, bearing long and dense terminal racemes of scarlet flowers. Stems and petioles are sometimes spiny.

Ficus spp.

A large range of mostly evergreen trees found in temperate to tropical regions. What is best with these trees is that these large trees lend themselves even to miniature bonsai. In matured bonsai, leaves are reduced in size to be proportionate with miniature trees. Some species are now available with small glossy green leaves and look very attractive. These will conform to almost every bonsai styles. But best be grown in its natural style of growing. Potting is possible at any time when the tree is not in its active growth. Wiring should be carried out in summer or during early rains.

Ficus benghalensis (Banyan tree)

A large spreading tree, often reaching a height of 30 m, gives off aerial roots which become thick after penetrating the ground. In due course, these can be trained into multi-trunk style. Large, leathery leaves, 10-20 cm long, fall in winter and new leaves appear in spring. It is an easily available plant, and can be selected from naturally grown seedlings. Large poll cutting can also be used for propagation of this species. In bonsai, the method is very much used for getting a tree with good trunk diameter within short time.

Mature tree gets reduced leaf size. Off time defoliation of leaves results in considerable reduction in leaf size. Bright crimson berrics ripen in May-June and may continue for a long time. Very easy growing and needs no special soil but good drainage is always encouraged.

Ficus krishnae (Butter cup tree)

A very curious form of *F.benghalensis* with typical irregular cup-shaped leathery leaves. A small tree seldom exceeds 5 m in height. The plant is usually propagated through layering.

Ficus benjamina (Java fig)

Large, graceful, evergreen tree reaching 15-20 m in height with gracefully drooping branches and bright, glossy light green, small, elliptical leaves. The tree also forms aerial roots; bark smooth, grey. It gives globular or ovoid fruits among its leaves, which turn reddish when ripe. The tree is also propagated from seeds or cuttings. Thick stem also respond to cutting. Bonsai of this species can be used for indoor decoration also.

F.benjamina has few more attractive forms with variation of size and shape of the leaves.

Ficus infectoria

An indigenous tree attaining a height of 15-18 m. Short trunk often knobby and irregular; pale, grayish, smooth bark, also gives off aerial roots. A deciduous tree, leaves fall in winter and reappear in January or

March. Leaves are glossy green. Young leaves which appear in March are copper coloured. Large branches of this tree can also be used for propagation.

Ficus religiosa **(Peepal tree)**

A very common tree used in bonsai. A large tree, 15-20 m in height, favoured by the bonsai artists because of its erect trunk with large vertical ridges, knobs, hollows and crooked branches with almost horizontally spreading. The tree becomes leafless for a short period in winter and again becomes covered with pinkish leaves in spring. The tree is often found growing in nature through seeds and young plant can be collected for future bonsai.

Ficus rumphi

A well-shaped tree, widely seen in nature with smooth, shining leaves of wavy edges. The tree bears fruits in pairs, black when ripe but whitish with black spots when young.

Ficus diversifolia

A tall shrub with spreading branches. Leaves obovate, thick, leathery, dark green with brown specks on the upper surface and pale on the lower side. It produces small fruits on all the branches, yellowish in colour. The plant grows well in shade and can be used for indoor decoration. Propagated by cutting or by air-layering. The species is suitable for training into any style of bonsai.

Ficus nitida **(Indian laurel)**

A very attractive looking, well-shaped plant with dense foliage and buttressed trunk. It extends aerial roots easily. Small elliptical deep green leaves, with very smooth waxy texture. It tends to be trained in any style of bonsai. Propagation is through cutting or layering.

Ficus triangularis **(Butter cup tree)**

A small tree, well-suited for pots. Leaves triangular, thick, dark green on the upper surface and pale underneath. It flourishes well in semi shade producing small reddish yellow figs profusely. Propagation is mainly by cutting and layering. This species is very suitable for a handsome upright style of bonsai.

Ficus tsjakela

A large spreading tree, often reaching a height of 15-17 m. The trunk is often knobby and irregular, pale, grayish. Leaves are thick, leathery and dark green. The plant grows well in shade. The species is suitable for training into any style of bonsai.

***Fortunella japonica* (Kumquat)**

An attractive tall, fruiting large shrub having scented flowers in summer and rains. It is suggested to grow this as a medium-sized or large bonsai, which in turn will give abundant edible fruits. Bonsai style includes informal, slanting, twin trunk and clump style. Wiring and pruning should be done in late spring. Propagation is by cuttings or air-layering.

***Jacaranda mimosaefolia* (Jacaranda)**

A deciduous, elegant tree with open branching system. The foliage is fern-like and bipinnate. The bell shaped blue - massive flowers are borne in erect panicles during March -May and just at the start of flowering the tree remains leafless for a brief period. It is propagated from seeds and semi hardwood cuttings. Informal upright form is the common style.

Juniperus spp.

A group of popular evergreen conifers for bonsai culture. Juniper, under the genus *Juniperus* has several suitable species for bonsai. Some are trees and some are low growing, often with thin flaking bark. A hardy plant, requires plenty of sunlight, responds well to pruning. Leaves bluish green, needle-shaped, mostly in whorls of three, or in pairs, some species with flat and scale-like leaves.

Potting can be carried out almost throughout the year, and it may be needed every two or three years. Wiring also can be done anytime of the year. Pruning at younger stage is advisable; nipping out the new buds leaving a few needles is also done to maintain the bushy and compact shape. Propagation through seeds and cuttings. Juniper is a slow-growing plant and can be trained into informal, slanting, semi cascade or cascade style.

Juniperus barbadensis

Densely branched, pyramidal tree, bluish green, sharply pointed needle-like leaves on young trees and scale-like as the tree ages. Not hardy, requires frequent watering.

***Juniperus chinensis* (Chinese juniper)**

Many varieties of the tree are found, but the variety 'Pyramidalis' is commonly used in tropical bonsai. A densely growing small tree or large shrub with bluish green, needle-shaped leaves mostly in whorls of three. The plant is very susceptible to pests.

***Juniperus horizontalis* (Creeping juniper)**

A prostrate growing bush with ascending branchlets raising very little above the ground foming a carpet-like mass of foliage. This plant is suitable for semi cascade or cascade style.

***Kigelia pinnata* (Sausage tree)**

A medium to large-sized evergreen and spreading tree having a globular appearance and with a dense branching system. The cup-shaped, mottled, dark purplish flowers from different parts of the branches in long stalks giving the appearance of candelabra. The fruits are sausage shaped. It can be trained into formal and informal styles.

***Melia azadirach* (Persian lilac)**

A native of India, medium sized fast growing, deciduous tree similar in appearance to neem. The leaves are beautiful and fern like and the sweet scented lavender-coloured flowers appear in March - April. The round yellow fruits borne in clusters also look ornamental. For bonsai, the plant can be styled as formal, informal and multi trunk style.

Malpighia coccigera

A slow-growing dwarf bushy shrub, with glossy green ovate leaves. Flowers pinkish white during summer and rains; fruits red. For bonsai, the plant can be styled as formal, informal, broom and clump styles.

Malpighia glabra **(Barbados cherry)**

This is a slow-growing, evergreen large shrub with small shining dark green leaves. Flowers are rose-red with trill petals. The advantages of this plant are the fruits which are cherry-like, showy, yellow, changing to scarlet. Propagation by seeds or by cuttings.

Milletia ovalifolia

A dwarf tree with ornamental ovate- shaped leaflets. It flowers in leafless condition during April. The small lilac coloured flowers completely cover the tree. Propagation is from seeds. It can be trained into formal and informal styles.

Millingtonia hortensis **(Indian cork tree, the jasmine tree)**

A quick growing, straight, tall tree with large foliage divided into leaflets. The foliage is handsome. The branches are drooping. The trumpet-shaped, silver-white flowers are sweet-scented at night. Propagation is from seeds and root suckers. This species is suitable for a handsome upright style of bonsai.

Mimusops elengi **(Spanish cherry)**

A native of India, medium sized evergreen tree with a dense crown of dark green shining foliage. The white, very fragrant, drooping flowers are borne in clusters during April - July and September - November. It is best suitable for its natural style of growing.

Punica granatum **(Pomegranate)**

Pomegranate responds quickly to bonsai culture with fruits. It is a tall, deciduous, spiny shrub or small tree with shining, oblong, narrow leaves. Flowers single or double, scarlet or orange scarlet in summer and rains. Single flowering types shall be selected for 'bonsai with fruits'.

The plant is propagated through cutting or layering. It can be trained into formal, informal, slanting or into a multi-trunk style.

Punica granatum **'Nana'**

A dwarf plant about 50 cm in height, bears small scarlet double flowers and small fruits. It is very suitable for miniature bonsai of many styles.

Pinus roxburghii **(Indian pine, chir)**

Pines are, undoubtedly, one of the best plants for bonsai. This evergreen conifer responds to almost all bonsai styling, but perhaps the most outstanding shape of all is the formal upright. Pruning for this plant is somewhat different. When the new shoots come out, they are pinched back, leaving a new bunch of needle-like leaves. This encourages new shoots and those can be treated in the same way. Branches which have been allowed to grow can be pruned back to five clusters of needles. Pruning of branches shall be repeated once in two years and they produce dense clusters of leaf bundles creating a very natural appearance. Repotting is done every two to three years, preferably in early rains. Wiring can be done at any time of the year. Pines are sensitive to water logging. Precaution should be taken while preparing soil mixture.

Salix babylonica **(Weeping willow)**

A broad-headed, medium-sized, deciduous tree, very graceful in appearance, with pendulous branches, usually planted around water surface. Of Chinese origin, its bright, hanging leaves attract bonsai growers. Its appearance suggests weeping branch style as the most suitable, or it can also be used as a rock setting bonsai. Propagation is easy from cuttings.

Tamarindus indica **(Tamarind)**

An attractive tree of tropical African origin which can be used freely in bonsai. It is a hardy tree and can tolerate severe pruning. It is easily grown from seeds.

Thespesia populnea (Portia)

A medium sized, quick growing evergreen tree. The leaves are poplar like and form into a head of foliage. The large bell shaped flowers are yellow in colour with reddish blotches at the base. Propagated from seeds or cuttings. The plants are salt resistant and hardy.

10

Trees for Renewable Energy

Renewable energy sources, are the sources of power that quickly replenish and which can be used again and again. For this reason they are sometimes called energy resources. The main sources of renewable energy and their advantages and disadvantages are :

Solar

Energy from sunlight is captured in solar panels and converted into electricity. Potentially it is an infinite source of energy supply, but the solar panels are costly.

Wind

Wind turbines (modern windmills) turn wind energy into electricity. This also is an infinite source of energy supply, but costly.

Tidal

The movement of sea water in and out drives turbines. A tidal barrage (a kind of dam) is built across estuaries, forcing water through gaps. In future underwater turbines may be possible out at sea and without dam. It has the potential to generate a lot of energy this way. But it is not eco-friendly.

Waves

The movement of sea water in and out of a cavity on the shore compresses trapped air, driving a turbine. It should be ideal for an island country.

Geothermal

It is possible to use the heat under the earth in volcanic regions. Cold water is pumped into earth and comes out as steam. Steam can be used for heating or to power turbines creating electricity. Potentially infinite energy supply. It is used successfully in some countries, such as New Zealand. It is dangerous and has to be done carefully.

Hydrological or Hydroelectric Power (HEP)

Energy is harnessed from the movement of water through rivers, lakes and dams. This creates water reserves as well as energy supplies. It is costly.

Biomass

This is decaying plant or animal waste. It is a cheap and readily available source of energy. If replaced, biomass can be a long-term, sustainable energy source.

Wood

Obtained from felling trees, burnt to generate heat and light. A cheap and readily available source of energy. If the trees are replaced, wood burning can be a long-term, sustainable energy source. If trees are replanted this is a perennial resource.

Greater reliance on renewable energy sources offers enormous economic, social and environmental benefits. It has been realised that, unlike conventional fossil fuel, the renewable or non-conventional energy resources cannot be put to use directly (except when they are used as fuel for household use). They need transformation into a suitable form of energy or the equipment needs to be changed to make use of them.

Biomass as a source of renewable energy

Biomass is organic material made from plants and animals. Biomass contains stored energy from the sun. Plants absorb the sun's energy during photosynthesis. The chemical energy in plants gets passed on to animals and people that eat them. Biomass is a renewable energy source because we can always grow more trees and crops, and waste will always exist. Some examples of biomass fuels are wood, crops, manure, and some garbage.

When burnt, the chemical energy in biomass is released as heat. Burning biomass is not the only way to release its energy. Biomass can be converted to other usable forms of energy like methane gas or transportation fuels like ethanol and biodiesel. Ethanol is an alcohol fuel made from the sugars found in grains, such as corn, sorghum, and wheat, as well as potato skins, rice, sugar cane, sugar beets, and yard clippings. Scientists are working on cheaper ways to make ethanol by using all parts of plants and trees. Ethanol is proving to be a very effective option for reducing our dependence on petroleum - regardless of whether it is made from corn or cellulosic materials. There is little doubt that ethanol will be the fruit biofuel that we can use to reduce our dependence on petroleum.

Biodiesel is much less polluting than petroleum diesel. It results in much lower emissions of almost every pollutant: carbon dioxide, sulfur oxide, carbon monoxide, air toxics and unburned hydrocarbons. Biodiesel does have nitrogen oxide emissions that are about 10 percent higher though. Blending biodiesel into petroleum diesel can help reduce emissions. Biodiesel contains almost no sulfur and can help reduce sulfur in diesel fuel used throughout the country.

Using biomass for energy can cut back on waste and support agricultural products grown in the country. Generating energy through biomass crops offer a number of environmental security benefits.

- Greater energy independence as a nation
- Opportunities for rural development and economic growth.
- Decreased dependency on non-renewable fossil fuels

Biomass and the environment

Biomass can pollute the air when it is burnt, though not as much as fossil fuels. Burning biomass fuels does not produce pollutants like sulfur, that can cause acid rain. When burned, biomass does release carbon dioxide, a greenhouse gas. But when biomass crops are grown, a nearly equivalent amount of carbon dioxide is captured through photosynthesis. Purpose grown trees as biofuel resources offer numerous advantages in productivity, sustainability and environmental impact. Sustainable forests provide habitat for wildlife and help clean the air by removing CO_2 and replenishing O_2 supply. Tree crops help minimize erosion and create a production tree for land unsuitable for other agricultural trees.

Biomass energy and global warming

Since plants and trees absorb and store atmospheric carbon as they grow, growing and using biomass energy crops reduce the level of CO_2 emissions into the atmosphere - which may be creating global warming and climate change on our planet.

The science behind this strategy to reduce greenhouse gas levels is accomplished in two ways. First, biomass energy from crops is "carbon cycle neutral" just like other forms of renewable energy such as wind or solar power. Second, growing energy crops creates a "carbon sink" through terrestrial carbon sequestration by storing carbon underground through root systems and soil chemistry management practices (e.g. recycling bagasse). Because of this creation of a "carbon sink" (a component which solar and wind energy do not have), we believe that bioenergy from closed loop energy crops represents the most effective choice in "alternative energy" options to address global warming.

Also it is important to note that biomass research and commercial demonstration is using environmentally damaged lands, such as closed mining sites. According to NASA Scientists, one-fifth of the carbon dioxide released annually from fossil-fuel emissions could be "sequestered" by planting energy crops on marginal lands of this type. Hopefully, this work can help create a "Global Model", where thousands of acres now largely considered wastelands can have productive agriculture and environmental use.

For marginal lands such as mining (phosphate, coal,etc.), pre-mined lands were most likely in native forest for hundreds/thousands of years. As such, these sites' soils were probably at carbon saturation. After mining however, empirical research is clear that post-mined lands often have little soil carbon.

Biomass power and Bio-fuels for rural employment and prosperity

In countries like India with rising energy requirements and dependence on imports of fossil fuel, considerable stress is being placed on developing renewable energy sources. India has approximately 50 million hectares of degraded wasteland and another 34 million hectares of protected forest area where tree cover is severely degraded. A massive programme to develop energy plantation of fast-growing tree crops such as Bamboo, *Casuarina* and *Eucalyptus* can serve as the model for a national network of small, decentralized biomass power plants. These power plants size from 10-25 MW can generate thousands of megawatts of power from renewable energy sources in a cost-effective manner.

India also has the capacity to generate biofuels in massive quantities. *Jatropha* plant, introduced from Africa, already grows wild in India. The plant produces large seeds which contain up to 35 per cent oil that is a substitute for No.2 diesel and kerosene blended in diesel motor fuels up to 15 per cent. Ethanol, which can be produced from maize, tapioca, sugarcane and sugar beet has enormous potential. It can be mixed as a pollution-free blend with petrol and Ethanol-petrol fuel blends are utilised in more than 20 countries.

The greatest advantage of biomass power and bio-fuels is that they can generate rural jobs and stimulate enormous growth of rural income, especially among the weakened. Therefore, these strategies should not be regarded from the narrow perspective of energy from the wider perspective of national development.

The National Commission for Agriculture, appointed in the 1980's to look into agricultural production as well as related aspects like availability of fuel and requirement for energy, had already pointed out the need to take up an intensive programme of augmenting renewable energy resources like biomass. The Government of India subsequently established a Department of Non-conventional Energy Sources (DNES) and also Energy Development Agencies in each state. These organisations identify suitable approaches for different states and promote research and development activities to enable the proper utilisation of renewable energy sources.

Energy crops- sustainable renewable energy source

An important aspect of "Energy Crops" is that they can represent a sustainable renewable energy resource - since our trees and certain crops like sugarcane will re-grow after each harvest (coppice, ratoon) - allowing multiple harvests without having to re-plant (called short rotation crops). The other functional uses of these crops are the following.

- A clean alternative energy biofuel for power plants.
- Feedstocks for ethanol (cellulose or sugar/starch platforms) and biodiesel.
- New cash crops to farmers & rural economic development.
- Greater Energy Independence (foreign oil & natural gas).
- Productive use of environmentally damaged lands.
- Natural wildlife habitats are preserved and enhanced.
- Energy crop trees used as "Bridge Crops" to reclaim/restore severely damaged closed mining sites.

Multipurpose trees in integrated farming systems

Multi-purpose trees and water plants supply the protein and the trees play other important roles such as controlling erosion, providing sinks for carbon dioxide (the standing biomass) and methane (at the interface between the decaying fallen leaves and the soil) and as a source of biodiversity. Sugar cane and trees have well developed systems of biological pest control, require minimum synthetic chemical inputs and are easily separated into high and low fibre fractions as required for the different end uses of feed for monogastric and ruminant animals and fuel. The trees which have had most farmer impact are *Gliricidia sepium, Trichantera gigantea, Erythrina glauca* and *edulis*. The foliage from *Trichantera gigantea* is consumed by pigs; that from the other trees is more appropriate for ruminants. The uses and advantages of multipurpose trees arc the following.

- Sources of protein for pigs, ducks, sheep and buffaloes (the leaves)
- Substrate for gasification (the trunks and branches)
- As sinks for CO_2 (the standing biomass)
- Sinks for CH_4 (the decaying organic matter from fallen leaves and unused branches)
- As media for synthesis of ammonia through the action of symbiotic microorganisms attached to, or in free-living associations with, the root layer.
- To control erosion
- To provide construction material and firewood
- The characteristics of multipurpose trees that are sought by farmers relate to the need to minimize inputs and maximise outputs. They require:
- High yield
- Ease of harvesting
- Resistance to pests and diseases
- Animal response to minimal inputs
- Ease of establishment and management
- Availability of different species

Six species are presently being used as multipurpose crops for integrated farming

- *Gliricidia sepium*
- *Trichantera gigantea*

- *Erythrina glauca*
- *Erythrina edulis*
- *Acacia mangium*
- *Prosopis juliflora*

The choice of a particular tree will be decided by climatic and soil considerations. Thus *Erythrina glauca* is specifically adapted to acid soils with a high water table, and a propensity to flooding. *Acacia mangium* is especially tolerant of acid soils saturated with aluminum. *Leucaena leucocephala* is not normally used as it is least tolerant of all the species to acid soils. Where *Leucaena* can be grown, it is out yielded by *Gliricidia,* which is also more resistant to insect attack and is more easily harvested. The latter is important in "cut and carry" systems which is the usual practice on small farms.

High-tech trees

For biofuels to become mainstream fuels, they must become economically competitive with petroleum, liquid that quenches most vehicles' thirst fairly inexpensively. Farmers, therefore, must start with extremely hardy trees remarkably resistant to drought, insects, and diseases that can produce high yields in many kinds of cropland. That combination of attributes isn't normally found in nature, so researchers must find ways to nurture it in the laboratory.

Using the same *gene-mapping* tools developed to unlock the secrets of human DNA, plant geneticists in Oak Ridge and at the University of Washington are probing the genes of fast-growing poplars. By finding the genes responsible for traits such as height, drought tolerance, and disease resistance, researchers can identify individual trees with particular promise, cross them with other trees which excel in other traits, and then produce genetically identical copies After screening dozens of species of trees for growth rates, adaptability, drought tolerance, and disease resistance. Researchers have found particular promise in the members of the genus *Populus*-which includes poplars and the willow.

The architecture of a tree is more than just aesthetic. The shape of a tree's crown, the geometry of branches and the size and spacing of leaves determine its efficiency to capture solar energy. The crown of the hybrid poplar combines all these qualities and the biomass production is as high as 10 tons of dry biomass per acre, per year. This is about 5 to 10 times more than the tree produces in the wild habitat. Taking the best individual hybrids and crossing them with other superior specimens, trees with even faster growth rates as well as greater drought tolerance and insect resistance can be produced.

Biomass crops could help revitalize farm economy by utilizing idle farm land. By stabilizing soil, reducing erosion and runoff and restoring organic matter to depleted land, these crops could restore the vigour of the land. They could boost the environment diversity of the farm belt by providing shelter for birds and other wildlife.

Functional classification of energy crops

Prior to the Industrial revolution, a much smaller world and population and consequentially limited land and resource consumption meant that the earth's resources and its ability to absorb pollution were never in question. In the last decade sustainability has become a buzzword in the global scenario. Sustainable development is development that meets the present without compramising the ability of future generations to meet their own need. Energy crisis is identified as one of the important problems facing our country. Energy crops which serve as sustainable energy resources hold promise for a better future. Their functions as renewable energy sources are many as given below.

Vegetable oils

The use of vegetable oils as fuel is gaining importance because of the escalation in petroleum prices to prohibitive limits. Research is in progress all over the world to use vegetable oils from a variety of plants and mix it with diesel in varying proportions for use as a fuel in vehicles with diesel engines. Therefore, farmers in the developed countries are encouraged to cultivate edible oil yielding plants.

Oils from a number of plants like groundnut, sunflower, safflower, maize, soybean, olive and palm have been found to be useful for this purpose. Besides, castor, rapeseed, eucalyptus, tung (*Aleurites sp.*) squash (*Cucurbita sp.*) malmeleire (*Croton sanderianus*) and babassu nut (*Orbignya sp.*) are also being tried for their oil.

The chief drawback of using vegetable oils as fuel is the cost factor which has been examined in various countries. The amount of oil extracted depends upon the productivity of the crop and ease of extraction. Trials with sunflower and soybean in South Africa, USA and Brazil have shown that the yield is 1 t/ ha. which accounts for 40-50 per cent of the weight of the harvested crop. With improved cultural practices the yield can be increased to 2 to 5 tonnes per hectare. In India there is an acute shortage of edible oils. Coconut palm and oil palm are potential edible oil plants for Indian conditions.

Non-edible oils

There are a number of non-edible oil seeds belonging to *Meliaceae, Sapotaceae, Guttiferae, Lauraceae,* and *Fabaceae.* Many of the species of shrubs and trees concerned contain 20 to 70 per cent fat. Some of the common non-edible oil seeds are neem (*Azadirachta indica*), mahua (*Madhuca indica*), sal (*Shorea robusta*), karanj (*Pongamia pinnata*), nahor (*Mesua ferrea*), kokum (*Garcinia indica*) and undi (*Calophyllum inophyllum*). The oils have been refined before use for industrial purposes. No attempt has been made for using these as an energy source in commercial scale. However, if these are cost competitive and available in sizeable quantities with sustained supply, their use for the purpose of energy may be explored either in the crude form or after processing.

Liquid wax

In comparison to lubricants from fossil, the liquid wax has excellent lubricant properties. Jojoba (*Simmondsia chinensis*) a monotypic dioecious species belonging to *Simmondiaceae*, has become important for the production of liquid wax since 1970.

Jojoba is now being cultivated on a large scale in the US, Mexico, Israel and Sudan. The tree grows well in marginal soils. It stands salinity, requires little water and does not suffer from any major pest or disease. Its cultivation requires minimal care requiring no special equipment. The oil can be expelled from seeds as easily as from any other oil seed. The tree is evergreen, non-poisonous, drought-resistant deep-rooted long lived and requires low maintenance. Moreover, the species is useful in landscaping highways and can reduce particulate pollution. It is also useful in soil conservation and stabilization.

Hydrocarbon plants

Petroleum is a hydrocarbon and is essentially a product of photosynthesis. Obviously time has come when plants that produce hydrocarbon of a suitable weight need to be investigated. *Hevea brasiliensis*, the tree yielding rubber, is very popular. However, there are other species of which have latex with molecular weight of 10,000 and are more relevant. A search for other renewable sources of natural rubber has shown that Guayule (*Parthenium argentatum*) and Andelion (*Taraxacum koksaghyz*) can be used for this purpose.

Petroleum plants

The hydrocarbons from plants which have a molecular weight of about 10,000 are suitable for further cracking to gasoline materials. Such selected plants are grown for the purpose of energy either for direct use or as feedstock for more convenient liquid fuel or other energy chemicals. These plants may be either used to get diesel fuel or after their conversion, for high quality liquid fuel. Recently Jatropha is identified as a biodiesel plant. Some of the species of *Euphorbia* appear to be another promising material. These plants need to be cultivated and subjected to an intensive investigation.

There are two oil yielding species which have recently been known as *Capaifera langsdorfii* and *C. multiguga*. The oil from *C. longsdorfii* is a perfect substitute for diesel in engines without any modification. The consumption is a little higher than that of regular gasoline. Otherwise its performance appears to be the same.

Description of energy crops

Scientific name : *Aleurites fordii*

Common name : China wood-oil/ Tung-oil tree

Family : *Euphorbiaceae*

Native of China, now mostly grown in the tea estates of Kangra Valley in Himachal Pradesh and in Assam and Bengal to provide shade for tea bushes. A high-quality drying oil, obtained from seeds, is used for preparing varnishes and paints. Oil dries rapidly to a film that is hard, water-proof and highly resistant to acids and alkalies. Tung-oil is utilized in the manufacture of industrial varnishes, for lithographic printing and aeroplanes. Other uses for tung-oil are in preparing quick-drying enamels, synthetic resins, for use in the textile industry and wall boards.

Scientific name : *Aleurites moluccana*

Common name : Candlenut/ Varnish tree

Family : *Euphorbiaceae*

An evergreen, wide spreading shade tree, native of Malaysia and the Pacific islands but grows wild in S. India and Assam. Oil, obtained from seeds is used in manufacture of paints and varnishes. Oil after drying makes a film that is hard, water-proof and highly resistant to acids and alkalies. Bark is used as tan. Seed-oil is used as purgative, and substitute for cator-oil. A compatible oil extracted from this is used for cooking food in Sri Lanka.

Scientific name : *Azadirachta indica*

Common name : Margosa tree/ Neem tree

Family : *Meliaceae*

It is a large and handsome tree, upto 18 m high with almost a straight trunk.

Leaves: Leaves are imparipinnate, crowded at the ends of branches, 22.0-40.0 cm long and glabrous; leaflets are 5-15, opposite, subopposite or alternate, lanceolate or ovate-lanceolate, acuminate, serrate or dentate and glabrous.

Flowers: The flowers are scented and are borne in numerous axillary panicles and pedicels, which are shorter than the leaves.

Fruits: Drupes, the fruits are 1-seeded, oblong and greenish yellow, when ripe. Flowering is in February-April and fruiting in June-August.

Tree is considered a good purifier of air, due to its large leaf area. Due to its bitter taste and disagreeable odour, not removed by conventional methods, neam oil has not been utilized on an industrial scale. However, a process has recently been developed to extract the bitter constituents from neem-oil. Oil thus obtained has been found to be suitable for soap-making and for hydrogenation. Seed-oil is also used as antiseptic and for burning purposes.

Scientific name : *Bambusa striata*

Common Name: Bamboo

Family : *Poaceae*

Native to China and Japan, commonly cultivated all over India.

A tufted bamboo, culms up to 8 m high, 5-8 cm in diameter, glabrous, yellow with green stripes, shining internodes 10-15 cm long, culm-sheath 15-20 cm long.

Leaves: Leaves are linear-oblong or oblong-lanceolate, 15-20 cm long, 2-2.5 cm wide, base obtuse, glabrous, glaucous beneath, leaf-sheath smooth, glabrous, ligule short, ciliate.

Flowers: Flowering is a rare occurrence.

It is mainly grown for its ornamental value; culms are used for construction and for paper pulp.

Scientific name : *Bambusa tulda*

Common Name : Bamboo

Family : *Poaceae*

A semi-deciduous, tufted fast growing erect bamboo. Culms 12-20 m high, 5-10 cm in diameter, glabrous, grey-green on maturity, internodes 30-50 cm long, culm-sheaths 15-20 cm long and broad, deciduous, blade broadly triangular or cordate, erect, hairy within.

Leaves: Leaves linear-lanceolate or lanceolate, 15-20 cm long, acuminate, base rounded.

Flowers : Flowering was reported from Bengal, Assam, Chittagong in Bangladesh and Myanmar. Inflorescence ranging from large leafless panicle to short leafy spicate branch, bearing clusters of few usually fertile long spikelets, spikelets 2.5-7 cm long, cylindrical, acute, with 4-6 fertile flowers and 1-2 male or imperfect flowers.

Fruits: Caryopsis, oblong, 7 mm long, hirsute at the apex. This bamboo is extensively used in North-East India for construction of huts, for making baskets, mats, food grain containers etc. It is suitable for production of pulp and paper. Tender shoots are edible, often used as vegetable and for making pickles.

Scientific name : *Calophyllum inophyllum*

Common name : Alexandrian laurel/ Dilo oil tree

Family : *Clusiaceae*

A medium-sized evergreen tree, upto 16m high, bark smooth grey or blackish brown.

Leaves: Obovate or ovate-oblong, coriaceous, 10-20 cm long, shining green, veins numerous, conspicuous underneath.

Flowers: White, scented, polygamous

Fruits: Spherical drupes, 2.5 cm in diameter, pulpy yellow when ripe.

Scientific name : *Casuarina equisetifolia*
Common name : Beef-wood, Casuarina, Australian oak
Family : *Fabaceae*

Equisetifolia means leaves like an 'equisetum' or horse tail. The tree is a native of Australia, Malaysia and the Pacific islands. It is a tall, straight stemmed tree reaching a height of 30 m.

Leaves: The ends of branches thickly set with numerous long, slender, usually drooping branchlets, mostly deciduous, performing the function of leaves; branchlets joined, 15-20 cm long, the internodes 4-6 mm long, 6-8 ribbed, from a distance branchlets look like the needles of pine.

Flowers: Pinkish, inconspicuous; male flowers appear in spikes at the end of branchlets, at the base of which the female heads (cones) are borne on comparatively thicker twigs; male spikes about 2 cm long; female cones, very short-stalked, globular, about 2 cm in diameter; ovary single-celled, with two ovules of which one develops into seeds; male and female flowers are usually borne on the same tree, sometimes on the same branch, in certain cases flowers of each sex are borne on separate plants.

Also planted as an avenue tree, and used for fuel. Tree is able to grow on sandy-coast near sea, used for reclaiming sand-dunes. Wood is used for transmission poles and for paper-making and bark is used as tan.

Scientific name : *Cocos nucifera*
Common Name : Coconut palm
Family : *Arecaceae*

Widespread and cultivated for centuries throughout the tropics, but believed to be of Indo-Malayan origin. This palm has a relatively slender, flexible stem which can grow to a height of 30-40 m marked in rings each year by the scars of fallen leaves and crowned by an apical cluster of huge long pinnate leaves.

Flowers: The flowers united in inflorescences with a spadix and bracted, are unisexual with the females below, generally solitary and formed by six twining sepals, while the numerous males grow at the top of each branch.

Fruits: The fruit is the well-known coconut, actually a big drupe containing a sweet liquid which becomes fleshy on ripening and when dried constitutes the copra used as a source of oil and fat. The whole plant is very important economically, both for its fruit and its wood; the shoots, too, are used for food. Fruit yields edible oil and wood is used for making charcoal.

Scientific name : *Croton sanderianus*
Common name : Croton
Family : *Euphorbiaceae*

An aromatic shrub or a small tree occurring in India produces a gum resin which is used in varnishes. Research is in progress to use its oil after mixing with diesel in varying proportions for use as fuel in vehicles with diesel engines.

Scientific name : *Croton aromaticus*
Family : *Euphorbiaceae*

An aromatic shrub or small tree occurring in Peninsular India, produces gum resin which is used in varnishes.

Scientific name : *Croton oblongifolius*
Common name : Chuka
Family : *Euphorbiaceae*

Distributed throughout India, seeds yield an oil which is used as purgative and also as an insecticide.

Scientific name : *Elaeis guineensis*
Common name : Oil palm
Family : *Arecaceae*

The stout trunk, ringed by the scars formed by fallen leaves, may grow to 20 m and is crowned by a big apical cluster of leaves 2-5 m long, with spiny stalk and a great many linear lanceolate, acuminate segments

Leaves: Leaves hang down, the inner ones are erect, so that the crown appears almost globose.

Flowers: The flowers are unisexual, carried in separate inflorescence; the male palms have a spiny tip, the females have spadices with rather big stalks, also spiny tipped, and appressate, so that when they ripen they form a dense, round mass of fruit with the spines sticking out of it.

Fruits: There may be as many as 200-300 fruits, drupes 4 cm long, with a red or blackish fleshy aril surrounding a very hard black nut with a white, soft kernel. An oil extracted from the fruit is used in the manufacture of foodstuffs and soap, and the whitish butter from the kernel is used locally in cooking.

Scientific name : *Eucalyptus globulus*
Common name : Blue gum
Family : *Myrtaceae*

An evergreen tree that can reach 40 m and more in height, with deciduous bark which comes off in long strips, leaving the trunk patterned in silver-gray.

Leaves: The young leaves are rounded, sessile and opposite, but on maturity they become lanceolate, stalked, alternate and aromatic.

Flowers: The flowers are solitary, sometimes growing two or three together, with short stalks; they consist of a calyx in the form of a truncated pyramid and an operculum formed from four petals joined together, which falls at maturity. When the operculum falls, the many stamens and the style emerge to give the flowers a feathery look. Flowering occurs between autumn and spring and is followed by the ripening of the fruits, which are angular capsules with a great number of very light seeds.

The wood is used mainly for cellulose. The oil could be used as a substitute or could be mixed with diesel in diesel engines.

Scientific name : *Garcinia indica*

Common name : Kokam butter tree/ Kokam

Family : *Clusiaceae*

A slender evergreen tree with drooping branches found in the tropical rain forests of Western ghats.

Leaves: Ovate or oblong lanceolate 7-9 cm long, 3-4.5.0 cm broad, dark green above and pale beneath.

Fruits: Globose or spherical 3-4.5.0 cm in diameter, dark purple when ripe enclosing 5-8 large seeds.

Kokam fruits are edible and a sweet syrup is prepared from it. The seeds yield a valuable edible fat known in commerce as 'kokam butter'. It is edible and also used in soap and candle manufacture, and suitable for ointments and other pharmaceutical preparations.

Scientific name : *Hevea brasiliensis*

Common name : Hevea rubber

Family : *Euphorbiaceae*

A tropical tree that reaches a height of 18 m and is the principal source of rubber, which is obtained by treatment of its coagulated latex.

Leaves: The leaves are alternate, with long stalks, composed of three lanceolate-acuminate leaflets.

Flowers: The plants are monoecious, with small flowers, pubescent and lacking petals, clustered in panicles.

Fruits: The fruit is a capsule.

The latex is collected by making an incision in the bark, which is light gray. On account of its elasticity, resilience tackiness and low-heat build up there has been an ever-increasing demand for the natural rubber which is indispensable for a number of articles like high performance tyres of automobiles, aeroplanes, etc. The hydrocarbon of *Hevea brasiliensis*has suitable molecular weight for petroleum production.

Scientific name : *Jatropha curcas*

Common name : Barbados - nut

Family : *Euphorbiaceae*

A shrub or small tree, native to tropical America and is now cultivated along the Coromandel Coast and in Travancore.

Leaves: Lightly 3-5 lobed

Flowers: Small, yellowish green

Seed-oil is used for manufacturing candles, soap and as lubricant and for illumination. Tender twigs are reported to be used for cleaning teeth.

Scientific name : *Madhuca longifolia*

Common name : Mahua tree, Mowra butter tree

Family : *Sapotaceae*

This is a large and deciduous tree, with about 17 m height. Bark is grey or blackish with shallow wrinkles and vertical cracks.

Leaves: leaves and petioles are pubescent or tomentose. Leaves are clustered at the ends of branches, elliptic or oblong-elliptic and shortly acuminate with cuneate base.

Flowers: The flowers are many at the ends of the branches and are borne on drooping pedicels.

Fruits: The fruits are ovoid, fleshy and green and about 2.5-5.0 cm long. Seeds are 1-4. Flowering is during April-May and fruiting in June-July.

Seed kernels yield mahua oil which is used in skin diseases, rheumatism and headache. Mahua cake possesses insecticidal and pesticidal properties. It is used for hairwash along with *shikakai* (*Acacia rugata*).

Scientific name : *Mesua ferrea*
Common name : Nahor
Family : *Clusiaceae*

It is medium-sized or large evergreen tree with short trunk, which often buttresses at the base and with dark-red heartwood.

Leaves: Leaves are coriaceous and lanceolate; upper side shining and lower covered with a white waxy powder.

Flowers: Flowers are white, large and solitary or in clusters and fragrant.

Fruits: The fruits are ovoid, 2.5-5.0 cm long with persistent calyx. Seeds are 1-4, shining, and dark brown with oily and fleshy cotyledons. Flowers in February-April and fruiting is during September-October. The flowers, seeds, fruits, leaves and bark of the plant are used medicinally. Seed oil could be used as a substitute or could be mixed with diesel in diesel engine.

Scientific name : *Olives*

Family : *Oleaceae*

Olea is a genus of some 35-40 species native to warm temperate or tropical regions from southern Europe (Mediterranean area) to Africa (the main center), southern Asia, eastern Australia and New Zealand. They are small to medium-sized trees or shrubs.

Leaves: The leaves are opposite, simple

Flowers: The flowers are white, mostly bisexual, borne in axillary panicles or fascicles.

Fruits: The fruit is a more or less ovoid drupe, dark blue to black when ripe.

The important species are *Olea europaea*, O.*europaea* subspecies *europaea* and *sylvestris*. Olives have been cultivated since prehistoric times, and constitute a crop of fundamental importance in the daily life, economics and sociology of the people of the Mediterranean countries, providing them with their main source of edible oil. In earlier times olive oil was also used for lighting as well as cooking.

Scientific name : *Pongamia pinnata* syn. *P. glabra*
Common name : Indian beech, Ponnga oil plant
Family : *Fabaceae*

Pinnata in Latin meaning 'feathered'. Common around the margin of inland water bodies and also in the tidal and beach forest of India, Sri Lanka, Malayan Archipelago, extending to the coasts of South China, the Fiji islands and tropical Australia.

A moderate sized, spreading, deciduous tree reaching 10-12 m with smooth, greyish-brown bark, a short trunk and rounded crown. Although *Pongamia* grows in all types of soils in tropical region, it prefers moist soil and warm humid climate.

Leaves: About 25 cm long, pinnate; leaflets shiny dark green, ovate or elliptic, 10-12 cm long and 6-8 cm broad. Tree becomes leafless in the early spring and flowering occurs along with the emergence of new leaves.

Flowers: Rose or lilac, about 1.2 cm long in short racemes in the axils of new twigs. Flowering starts at the age of 3-4 years. The ground below the tree becomes thickly covered with fallen flowers.

Fruits: Pods woody, non-dehiscent, rhomboid, 4-5 cm long, 2-3 cm broad containing usually 1-3 seeds.

Seed-oil is used for soap-making, as an illuminant, and is also used in treatment of skin disease and rheumatism.

Scientific name : *Prinsepia utilis*

Common name : Bhikal

Family : *Rosaceae*

A thorny shrub, found throughout Himalayas and Assam. Fatty oil from seeds is used for cooking and as an illuminant. It is also used for hydrogenation and soap-making.

Scientific name : *Salvadora oleoides*
Common name : Pilu
Family : *Salvadoraceae*

A shrub or small tree with short twisted trunk and drooping branches, distributed in the arid regions of Punjab, Rajasthan and western India.

Leaves: Opposite

Flowers : Hermaphrodite small or functionally unisexual on the branches in axillary or terminal panicles

Fruit: Drupes, globose, supported by the slightly enlarged calyx and are edible.

Seeds erect, globose and yields fatty oil. Seed-cake or meal is reported to be valuable as manure for tobacco.

Scientific name : *Schleichera oleosa*
Common name : Lac tree/ Honey tree/ Macassar oil tree
Family : *Sapindaceae*

A tree, found in the sub-Himalayan tract, Bihar, Punjab, Madhya Pradesh, southwards to Peninsular India, serves as host for the lac insect. The seed-oil is used as an illuminant, lubricant, for soap-making and in perfumery. Oil-cake makes good manure. Raw fruits are pickled, and ripe ones possess acidic taste and are edible. Young shoots are lopped for fodder. Powdered seeds are applied to ulcers of animals for removing maggots. Timber is reported to be used for oil and sugar mills, rice pounders, felloes and stocks of cart wheels, and other agricultural implements. Treated timber is suitable for cabinet-work, beams and rafters. It makes good fuel and charcoal.

Scientific name : *Shorea robusta*
Common Name : Sal tree
Family : *Dipterocarpaceae*

A large, gregarious species, reaching up to 30 m, trunk clean and straight in forest grown trees, crown conical or elongated when young, afterwards rounded, bark thick, rough, dark brown with deep longitudinal furrows.

Leaves: Leaves are broad-ovate or ovate-oblong, 10-25 cm long, more or less acuminate, base cordate or truncate, glabrous, shining deep green, pinkish when young, orange-yellow at senescence, stipules caducous.

Flowers: Flowers are yellowish on short pedicels in unilateral racemes arranged in compound panicles.

Fruits: Fruits are ovoid, 10-15 mm long, reddish to yellowish green in colour, indehiscent, one-seeded with 5 somewhat unequal 5-7 cm long, wing-like persistent sepals, seed ovoid.

Sal is used extensively and is one of the highly coveted constructional and domestic timbers of India. It is used as such for sleepers, beams, scantlings, rafter and floors, bridges, pit-props, carriages and agricultural implements. It is also a good fuel wood. The sal wood is reported to yield a pulp suitable for the manufacture of paper in admixture with bamboo pulp. Sal bark along with leaves and twigs is a promising tanning material. Seeds yield 19-20 per cent of a fatty oil (sal butter). It is suitable for soap making after blending with other softer oil and as a substitute for cocoa butter in the manufacture of chocolate.

An oleoresin called sal-dammar obtained on tapping trunk is used to caulk boats and ships, in paints, varnishes and as incense. It is an ingredient of *samagri* which is burnt in religious ceremonies. Reported to have been used for hardening softer waxes for use in preparation of shoe-polishes, carbon papers, typewriter ribbons. Sal-dammar also finds use as plastering medium for walls and roofs and as cementing material for plywood and asbestos sheets.

Scientific name : *Shorea roxburghii*

Common Name : Sal tree

Family : *Dipterocarpaceae*

A large tree, attaining a height of more than 30 m, bark grey, somewhat smooth. It is a native of southern Deccan districts of Tamil Nadu, Karnataka and Kerala in India,

Leaves: Leaves are elliptic or oblong to ovate-elliptic. 7-12 cm long, acute or retuse, more or less rounded base, margin undulate.

Flowers: Flowers are white or pale pink, about 2 cm in diameter, in drooping axillary panicles, calyx segments 5, glabrous, ovate, basally connate, margin ciliolate, petals 5, oblong, 1.5 cm long, villous, stamens and anthers 15, in rows with a long seta.

Fruits: Fruits are ovoid-obconical, about 2 cm long, with unequal enlarged calyx, wings oblong, 4-7 cm long, unequal, prominently nerved, seed solitary, obovoid, 7 mm long.

The tree is planted for the timber. The flowers appear in March-April and fruits ripen in June-July. New foliage appears in February. The wood is hard, much used for house building and for bridges, piles etc. It is a good fuel wood. The tree is a valuable host of the lac insects. Flowers yield an essential oil.

Scientific name : *Simarouba glauca*

Common Name : Simarouba/ aceituno/ paradise tree

Family : *Simaroubaceae*

A medium-sized tree, up to 15 m high, bark grey, smooth, indigenous to Cuba and introduced and widely grown in many tropical countries.

Leaves: Leaves are 20-40 cm long, leaflets 11-21, oblong or oblong-elliptic, 5-10 cm long, rounded, retuse or abruptly pointed at the apex, cuneate at base, glabrous, coriaceous, dark green above, pale or glaucous beneath.

Flowers: Panicles 10-30 cm long, widely branched, the staminate more copiously flowered than the pistillate, calyx 3-3.5 mm wide, lobes ovate to triangular, petals oblong to obovate, 4-6 mm long, rounded, pointed at apex.

Fruits: Drupes ovoid or oblong, 1.5-2 cm long, slightly oblique, scarlet or dark purple. The tree is fairly fast growing and it is not browsed by cattle. It flowers in March-April. It is grown for fuel wood, oil from kernels and for establishing vegetation in degraded land exposed to long spells of drought. The fruits are edible.

All parts of the oil tree are useful. Seeds contain 60-75% oil. Simarouba oil extracted from seeds is consistent in quality, can be easily refined, bleached and fractioned. High oleic acid fraction (59.10%) with low melting point (27^{0}C) can be used as cooking oil while fats containing palmito stearin fraction with high melting point can be used as cocoa butter substitutes in confectionary and bakery industries. Oleochemicals can be produced from oil and they are used for making surfactants, detergents, soaps, cosmetics, stabilizers and lubricants, grease, emulsifiers, paints, varnishes and candles. A new development is possibility to use simarouba oil as fuel for diesel engines.

Scientific name : *Simmondsia chinensis*
Common name : Jojoba
Family : *Simmondiaceae*

Jojoba is now cultivated on a large scale in the U.S, Mexico, Israel and Sudan and grows well in marginal soils. It is even grown as non-poisonous drought resistant deep rooled and long lived tree. The species is useful in landscaping highway and can stand particulate pollution and is useful in soil conservation and stabilization. It's oil can be expelled from seed. It could compile with other lubricants as also sperm whale oil. The liquid wax requires no refining for use in industrial - processes and has high viscosity index and very high flash and fibre points and likes sulpher more than sperm whale oil. It requires no addition of mineral oils for suphurization and finally the liquid wax does not change its viscosity on repeated heating and remains undamaged. At present its liquid wax is going in the lucrative market of cosmetics and pharmaceuticals but the outlook is that it will be profitable as a high grade lubricant because of its renewable source. After expelling the liquid wax the jojoba seed meal contains upto 30 per cent crude protein and is a potential animal feed ingredient.

Scientific name : *Trema orientalis*
Common name : Charcoal tree/ Indian nettle tree
Family : *Ulmaceae*

A small deciduous evergreen tree 8-10 m high with straight, spreading branches and thin greyish brown bark.

Leaves: Alternate, obliquely ovate - oblong or oblong -lanceated 7-12 cm long, glabrous above, white-tomentose below.

Flowers: Small drupe, 2-3 mm long, globose, blackish.

The tree is used as fire wood.

11

Trees in Astrology

Vedic literature indicates that the energy in the universe gets transmitted into the human body through star galaxies. The star galaxies of our universe are divided into 27 parts each signified by a particular star (*nakshatra*). Every human being is assigned to a particular star by virtue of his birth time. Each star is associated with a tree. The tree associated with his star gives the universal energy to the human being. Thus astrology gives us a chance to identify the tree which is auspicious to us, being associated with our birth star. When one plants the tree associated with his birth star near his dwelling, it will bring good luck and prosperity to him.

According to the *Vedas,* trees act as tools to transfer the energy from star galaxies to human body and mind. In our body the energy is transformed into pranic energy. There are five different types of pranic energy. If there is any imbalance among the different forms of pranic energy, the health and mind get disturbed. The right balance of these five energy levels can be achieved through proper utilization of trees, if we have knowledge of our birth star and the tree associated with that star. Doing *'pranayama'* near the tree associated with his birth star cures a sick person. Thus the imbalanced pranic energy which made him sick could be balanced through his birth star-tree.

Table 5. Trees associated with star signs

S. No	Star	Scientific name	Common name	Family	Salient characters
1	*Aswathy*	*Strychnos nux-vomica*	Nux-vomica	Loganiaceae	An evergreen glabrous tree flowering during March-April, flowers greenish white in corymbose cymes. Fruiting in winter season.
2	*Bharani*	*Phyllanthus emblica*	Indian Gooseberry	Euphorbiaceae	A moderate sized deciduous tree, flowers during February-May. Greenish yellow flowers are borne in axillary clusters. Fruiting in October-March.
3	*Karthika*	*Ficus glomerata*	Cluster fig	Moraceae	A deciduous tree with much leathery leaves. Plant bears crimson berries during August-September.
4	*Rohini*	*Syzigium cumini*	Jambolan plum	Myrtaceae	A tall spreading tree. Flowers in April-May and produces purple-black oblong roundish fruits during June-August.
5	*Makayiram*	*Acacia catechu*	Black cutch	Fabaceae	A small tree with spreading crown and feathery foliage. Bark blackish, flowers greenish white on long spikes. Flowering in May-August and fruiting in November-February.
6	*Thiruvathira*	*Diospyros crumenata*	*Karimaram*	Ebenaceae	An evergreen tree with dense deep green foliage.
7	*Punartham*	*Bambusa arundinaceae*	Spiny bamboo	Poaceae	A tall woody grass with open clumps and are arching in nature. Monocarpic

Contd...

Table 5. Contd...

8	*Pooyam*	*Ficus religiosa*	Sacred Peepal	Moraceae	A large to medium sized deciduous tree with spreading branches. Leaves broadly ovate or round and long petioled. Flowering during summer and fruiting in rainy season.
9	*Ayilyam*	*Mesua ferrea*	Iron-wood tree / Nagakesar	Clusiaceae	A moderate sized tree, highly ornamental, lanceolate, shiny leathery leaves. Young leaves deep red, turning to pink and then green. Flowers white highly scented. Flowering during February-March.
10	*Makam*	*Ficus benghalensis*	Banyan tree	Moraceae	A huge tree with a huge crown of foliage. Aerial roots originating from branches strike into the soil. Plant bears crimson berries during August-Spetember.
11	*Pooram*	*Butea monosperma*	Flame of the forest	Fabaceae	A small to medium sized deciduous tree with trifoliate leaves. Flowers bright, flamy, scarlet orange which turns olive green to black. Calyces with a shape resembling a parrot's beak. Flowering during April-May.
12	*Uthram*	*Ficus gibbosa*	*Itthi*	Moraceae	A medium sized evergreen tree with a dense crown and shining foliage. Flowers in March-April and fruiting during May-June.

Contd...

Table 5. Contd...

13	*Atham*	*Spondias pinnata*	Indian hog-plum	Anacardiaceae	A small aromatic tree flowering observed in February-March and fruiting in September-October.
14	*Chithira*	*Aegle marmelose*	Bael tree	Rutaceae	A small or medium sized deciduous and aromatic tree. Leaves are trifoliate. Flowers are greenish white and sweet scented. Flowering in April-May and fruiting during March-April in the next year.
15	*Chothi*	*Terminalia arjuna*	Arjun tree	Combretaceae	A large tree with smooth grey bark and drooping branchlets. Leaves oblong, or elliptic - oblong pale dull green above and pale brown beneath. Flowers are yellowish white and borne in shortly panicled spikes. Flowering is in April-May and fruiting during September-November.
16	*Visakham*	*Asclepias curassavica*	Blood flower	Asclepiadaceae	A small shrub having medicinal properties.
17	*Anizham*	*Mimusops elengi*	Spanish cherry	Sapotaceae	A medium sized handsome evergreen tree with globular crown and shiny green foliage. Flowers are star shaped and fragrant. Berries ovoid and orange when ripe. Flowering in March-April and fruiting in May-June.
18	*Triketta*	*Aporusa lindleyana*	Vetti	Euphorbiaceae	A medium sized tree with dense foliage.

Contd...

Table 5. Contd...

19	*Moolam*	*Vateria indica*	Vella pine	Dipterocarpaceae	A medium sized tree that produces white dammer (Vellakunthirikkam). Flowering in March-June and fruiting in July-August.
20	*Pooradam*	*Salix tetrasperma*	Indian willow	Salicaceae	A medium sized tree
21	*Uthradam*	*Artocarpus heterophylla*	Jack tree	Moraceae	A large evergreen tree indigenous to Western Ghats with dark green leathery leaves. Multiple fruits are edible. Flowers during October-December and fruits produced in March-June.
22	*Thiruvonam*	*Calotropis gigantia*	Madar	Asclepiadaceae	A perennial under - shrub with light green velvety leaves. Flowers in January-February and fruits are produced during March-April.
23	*Avittam*	*Prosopis cineraria*	Jand	Fabaceae	A medium sized spiny tree with drooping branches. Flowers creamy-white, arranged in axillary or terminal panicles. Flowers in March-April and fruits produced in September-November.
24	*Chathayam*	*Anthocephalus cadamba*	Kadam	Rubiaceae	A tall deciduous tree having straight trunk with smooth bark and a dense rounded crown of dark glossy foliage. Flowers small, scented and cream coloured. Flowering is in February and fruiting in June-August.

Contd...

Table 5. Contd...

25	*Pooruruttathi*	*Mangifera indica*	Mango tree	Anacardiaceae	A large evergreen tree with widely spreading branches and thick dark grey bark. Leaves are crowded at the end of branches. Flowers greenish yellow and odorous in terminal panicles. Flowering during March-April and fruiting in June-July.
26	*Uthrattathi*	*Borassus flabellifer*	Palmyra palm	Arecaceae	A fan leaved palm with a long, stout stem. Nuts are edible flowering is in March-April and fruiting in July-August.
27	*Revathi*	*Madhuca longifolia*	Mahua tree	Sapotaceae	A large, deciduous tree with pubescent leaves clustered at the end of branches. Flowers yellowish white, borne on drooping pedicels. Fruits ovoid, fleshy and green. Tree flowers during April-May and fruits produced in June-July.

Bibliography

Bose, T.K., Das, P and Maiti, G.G. 1998. Trees of the World. Vol.I. Regional Plant Resource Centre, Bhuvaneswar, Orissa, India.

Bose, T.K. and Chowdhury, B. 2004. Garden plants in colour: Trees - Tropical and subtropical. Bose, T.K., Chowdhury, B., Cameron, P.M., Maiti, G.G. and Maiti, R.G. Partha Sankar Basu (eds.), Naya Udyog, Kolkata 239p.

Dhiman, A.K. 2004. Ayurvedic Drug Plants. Daya Publishing House, New Delhi

Editore, A.M. 1982. The MacDonald Encyclopedia of Trees. MacDonald and Company, London.

Hora, B. 1981. The Oxford Encylopedia of Trees. Oxford University Press, Oxford, New York.

Ingels, J.E. 1978. Landscaping: Principles and Practices. Van Nostrand Reinhold Company, New York. 69p.

Kohli, R.K. 1996. Need and planning for avenue trees in cities - A Chandigarh experience. Khosla, P.K. (ed.) Ecofriendly Trees for Urban Beautification, ISTS, Solan, p.41.

Mani, 1974. Ecology and Biogeography in India. Dr. W Junk. b.v. publishers, The Hague, 773p.

Myers, 1990. Biodiversity Challenge: Expanded Hot Spots Analysis. The Environmentalist 10: 243-256.

Nadkarni, K.M. 1954. Indian Materia Medica. Popular Prakashan, Bombay.

Nowak, D.J. 1995. Air pollution removal by Chicago urban forest. Chicago's Urban Forests Ecosystem. Pherson, E.G., Nowak, D.J. and Routree, R.A.(eds.). Technical Report, No. 186: 65-85.

Ramkrishna, N. 2006. Biodiversity in Indian Scenerios, Daya Publishing House, New Delhi.

Randhawa, M.S. 1978. Ornamental Horticulture in India. Today and Tomorrow Printers and Publishers, New Delhi, p. 1-39.

Randhawa, M.S. 1983. Flowering Trees. National Book Trust, New Delhi. 207p.

Singh, M.P. 2004. Natural Resources and Renewable Energy. Daya Publishing House, New Delhi. 472 p.

Subramanyam, K. and Nayar, M.P. 1974. Vegetation and Phytogeography of the Western Ghats. Mani M.S (ed.). Ecology and Biogeography in India. Dr. W. Junk b.v. Publishers, The Hague, 178 – 196.

Surapala's Vrikshayurveda translated by Nalini, S. 1996. Agri-History Bulletion No. 1. Asian Agri-History Foundation, Hyderabad, India. 35-62.

The Mother, 2004. Flowers and their messages. Sri. Aurobindo Ashram Trust, Pondicherry, India, p. I-XV, 131-163.

Colour Plate Section

Alstonia scholaris

Amherstia nobilis

Bauhinia purpurea

Bombax malabarica

Couroupita guianensis

Cassia fistula

Cassia javanica

Delonix regia

Holarrhena antidysenterica

Lagerstroemia flos-reginae

Peltophorum ferrugineum

Plumeria acutifolia

Plumeria rubra

Pterospermum acerifolium

Saraca indica

Spathodea companulata

Index

A

B

D

E

F

G

H

I

J

K

N

O

P

Q

R

S

About the Series Editor

Prof. K.V. Peter

Prof. K V. Peter, basically a horticulturist, a plant breeder and a University teacher is an acknowledged research manager. As Vegetable Breeder, he developed bacterial wilt resistant varieties in tomato, eggplant and chilli. The variety Pant C1 which he developed along with late Dr. P.J.Mathai is resistant to viral leaf curl and thrips. He authored the bulletin "Statistics and computer programmes in biometric research" published by G.B.Pant University of Agriculture and Technology, Pantnagar in 1976. Prof. Peter established a spices biotechnology facility at Indian Institute of Spices Research, Calicut and nurtured an active team of scientists who worked out the protocols for rapid multiplication of 32 spices including black pepper and cardamom. *In vitro* conservation of spices germplasm consisting 36 species and 315 accessions are now accomplished, along with know-how for minimal growth methods. Prof. Peter provided effective managerial support to the Indian Institute of Spices Research (IISR), Calicut to possess the world's largest collection of black pepper and cardamom germplasm. The high production technologies developed by the team of scientists led by Prof. Peter could push up cardamom productivity from 70 Kg/ha in 1992-93 to 149 Kg/ha in 1997-98. Black pepper productivity also shot up from 268 Kg/ha during 1992-93 to 311 Kg/ha in 1995-96. The technology package for protected cultivation of bush black pepper would make available green pepper throughout the year. Prof. Peter edited 10 books and is author or co-author of 13 books/bulletins and 124 research papers. Prof. Peter edited 3 volumes of *'Handbook of Herbs and Spices'* published by *Woodhead Publishing Company,UK and CRC,USA*. Ha co-authored the *'Handbook of Industrial Crops'* published by *Haworth Press, USA*. The National Book Trust of India published the book Plantation Crops authored by him. He was Director of Research, Kerala Agricultural University, Vellanikkara providing research guidance and support to over 724 scientists. He is a Fellow of National Academy of Agricultural Sciences, New Delhi and Fellow of National Academy of Sciences, Allahabad. He was awarded the "Rafi Ahmed Kidwai Award" by Indian Council of Agricultural Research, New Delhi. The National Academy of Agricultural Sciences, New Delhi conferred 'Recognition Award 2000'. The University of Agricultural Sciences, Bangalore bestowed 'M. H. Marigowda Award' to him for outstanding contributions in Horticultural Sciences. He is the recipient of Silver Jubilee Medal for outstanding contributions in vegetable research, instituted by the Indian Society of Vegetable Science, Varanasi, UP. During 1991-2000, he got the then National Research Centre of Spices elevated into Indian Institute of Spices Research. Indian Council of Agricultural Research, New Delhi conferred The Best Institute Award on IISR for its research contributions to spices.Scientists of IISR were encouraged to receive national recognitions like Jawaharlal Nehru Award, Hari Om Trust Award, Best Team Research Award etc. during the period (1991-1998) Dr. Peter was Chairman, QRT, Indian Institute of Vegetable Research,Varanasi and All India Co-ordinated Vegetable Improvement Project (1998-2003), Member Secretary, QRT, Central Tuber Crops Research Institute, Trivandrum (1983-1988), Member, Standing Committee,Association of Indian Universities, New Delhi (2004-2005) and Member, Spices Board of India (1991-1998). He was Member, Governing Board, ICAR (2002-2005) and Member, Board of Management IARI, New Delhi (2003-2006). Prof. Peter is now Member, Task Force (Plant Biotechnology) DBT, UGC-DBT Human Resource Development, DST Task force on Women in Science and Chairman, RAC, National Research Centre for Medicinal and Aromatic Plants, Anand.

Born on 17th May, 1948 in the Island village of Kumbalangi, he had secondary school education at St. Peters High School, Kumbalangi, pre-degree at Maharajas College, Ernakulam, B. Sc. (Ag) at Agricultural College and Research Institute, Vellayani and Mastoral and Doctoral degrees at G.B. Pant University of Agriculture and Technology, Pantnagar and post doctoral training at USDA, BARC-W, Maryland. He attended scientists meet at Bangladesh, Taiwan, Goudeloupe (FWI) and USA. He is co-ordinator of two network projects on plant biotechnology, funded by Department of Biotechnology, Government of India.The Research Team led by him received the National Award For Processing and Product Development given by The Department Of Biotechnology.Under his Vice Chancellorship at Kerala Agricultural University, the University received 1st prize during 2001-2002, 2002-2003 and 2003-2004 and 3rd prize during 2004-2005 in academic performance on all India basis from ICAR, New Delhi. The ICAR New Delhi conferred. The Sardar Patel Award for the Best ICAR Institution 2003 to Kerala Agricultural University for allround performance. Four New Krishi Vigyan Kendras were established to serve farming community in 4 districts of Kerala.Dr Peter is Vice-Chancellor, Kerala Agricultural University since June, 2001.He guided several students for M.Sc and Ph.D programmes in the Universities of Calicut, Kerala Agricultural University and GBPUAT, Pantnagar.

Dr. Peter is the S/o Late Kuruppacharil Devasi Varkey, Kumbalangi. His wife is Mrs. Vimala Peter. He has two sons Mr. Anvar K. Peter, MCA and Mr. Ajay K. Peter, B. Tech. MBA student.

About the Authors

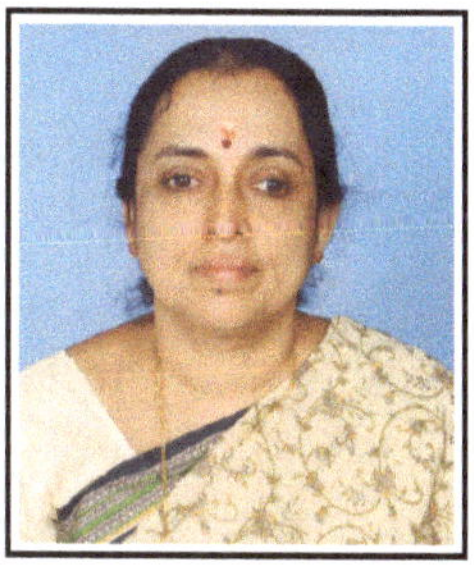

Dr.(Mrs.) P.K. Valsalakumari, Ph.D obtained her B.Sc.(Hort), M.Sc (Hort) and Ph.D from Kerala Agricultural University with distinction. She joined as Junior Instructor in the same University in 1977. At present she is holding charge of the Head of the Department of Pomology and Floriculture, College of Horticulture, Kerala Agricultural University. She has been associated with active research, teaching, and extension activities in fruits, flowers and landscaping. She has handled several courses in the areas of ornamental gardening, landscape horticulture and floriculture for the undergraduate and post graduate students. She has guided 10 students for their M.Sc and 2 students, for Ph.D. She has been associated with landscaping of the campus and consultancy services to public and private institutions in landscaping and floriculture.

Dr. Valsalakumari has 120 research papers to her credit including those presented in national and international conferences and seminars and 21 books/ chapters in books as editor, author or co-author and a number of training manuals and popular articles. She has implemented 8 externally funded research projects by ICAR, NHB, DBT and BARC and is the Co-Principal Investigator of several projects. She has played a major role in the organization of a National Seminar and the National symposium of the Indian Society of Ornamental Horticulture in Floriculture at the Kerala Agricultural University and a winter school on "Flori Business in the changing Global scenario: Issues and strategies" sponsored by the ICAR.

Dr. Valsalakumari has organized a number of training programmes in tree planting, ornamental gardening, nursery management, floriculture and cultivation of anthurium for the benefit of the officials of the Department of Agriculture and farmers. She has also established commercial nurseries and demonstration plots of anthurium and other flower crops in the University and in farmers' field for the transfer of technology.

Dr. Valsalakumari is the life member of Scientific societies like Indian Society of Ornamental Horticulture and is the referee of National journals. She is the associate Project Co-ordinator (Floriculture) and member of the Board of Studies of Kerala Agricultural University. She has served as external examiner of several universities. She has visited Thailand, Singapore and Malayasia to attend International Conferences and Shows. She served as judge in the International orchid show conducted in Malaysia in July,2007.

Dr. P.K. Rajeevan obtained his B.Sc. (Ag) and Ph.D. from Kerala Agricultural University and M.Sc. (Ag) in Horticulture from Andhra Pradesh Agricultural University. He joined the Kerala Agricultural University in 1978 as Assistant Professor and is presently serving as the Associate Dean, College of Horticulture and as Floriculturist of AICRP on Floriculture (Vellanikkara Centre). He has guided 8 M.Sc. and 3 Ph.D. students in Floriculture.

Dr. Rajeevan has been actively associated with research works on fruits and flowers. For the last 20 years he has been working on orchids, anthuriums, foliage plants and other tropical ornamentals and has made significant contributions in the areas of mass propagation, standardization of agro techniques, standardization of post harvest technology etc. He has operated various research and development projects funded by ICAR, DST and NHB. He has prepared projects for development of floriculture for the Government of Kerala and landscape projects for several Governmental organizations, including that for the international airport, Kochi.

He has visited Kenya, Saudi Arabia, Dubai, Mauritius, Singapore, Malaysia, Thailand and Sri Lanka on various assignments like representing India in International Orchid shows, offering consultancy service in floriculture, serving as a member of a committee constituted by the Government of Kerala supported by APEDA in Floriculture market exploration studies etc. He has participated in several national and international seminars and has also served as resource person in training programmes organized in various parts of the country including Lakshadweep and the Andamans. He is also acting as consultant for floriculture development for the Government of Goa. He has organized national symposium of the Indian Society of Ornamental Horticulture and a National seminar on Floricutlure (Flori 2020) at the Kerala Agricultural University. He has also conducted over 100 training programmes in floriculture and allied areas for the benefit of farmers and agricultural officers.

Dr. Rajeevan has served as member of the General Council (Senate) of Kerala Agricultural University, Joint Secretary of Indian Society of Ornamental Horticulture, External member of the Board of Studies in TNAU, Member QRT of NRC, Orchids and Councillor of The Orchid Society of India. He is now serving as Project Co-ordinator (Floriculture) of Kerala Agricultural University and Vice President of Indian Society of Ornamental Horticulture. He has published 141 research papers, 28 books/ chapters in books and 42 popular articles. He has also done TV and radio programmes on various aspects of floriculture and landscaping. He was awarded HSI Gold Medal for his contribution and leadership in Floriculture for the year 2005.

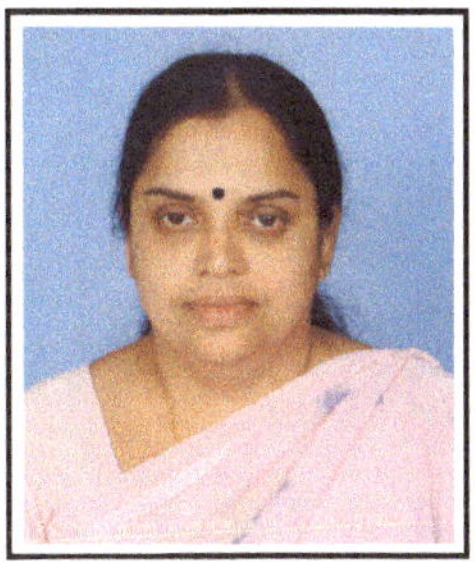

Dr. (Mrs.) P.K. Sudhadevi completed her Post graduation and doctorate degrees in Horticulture from Kerala Agricultural University with brilliant academic records. She has vast experience of 25 years in teaching, research and extension activities in Horticulture. She handled three externally aided projects as Principal Investigator and was associated with the implementation of three other schemes. She handled several projects on plantation crops, spices, medicinal, aromatic and fruit crops as Principal Investigator in KAU. Dr. Sudhadevi, is the Scientist who released the first pineapple hybrid in India 'Amritha' in association with other Scientists of KAU for which she was honoured by the Hon. Minister for Agriculture of the state. She attended several national and international seminars and presented papers. She has published over 55 research papers in national and international journals and symposia and chapters in 6 books and several popular articles. She also has 6 books published on her credit on different aspects of Horticulture.

Dr. Sudhadevi is life member of one International Scientific Society and several reputed bodies in India. She is presently working as Professor in the Department of Pomology and Floriculture of College of Horticulture, Kerala Agricultural University.

Dr. (Mrs.) C.K. Geetha completed her B.Sc and M.Sc degree and Ph.D. in Horticulture with First Rank, from the Kerala Agricultural University. She has vast experience of about 25 years of teaching, research and extension in Horticulture. She has been teaching undergraduate and post graduate students since 1986 and guided few Ph.D and M.Sc students. She has handled two externally aided schemes as Principal Investigator and 10 scheme as Co-investigator/ Associate, funded by various reputed bodies/ institutions. Dr. C.K. Geetha has published over hundred research papers in national and international journals and symposia, eight books, chapters in two books and several popular articles. She received the Certificate of Appreciation from the Vice Chancellor, KAU for operating the External Aided Projects in the University Dr. C.K. Geetha is life member of some reputed Indian Scientific Societies in the field of Horticulture and member in the Board of UG and PG Examinations in some of the reputed Agricultural Universities in India. She participated as a judge in the 7th Ipoh International Orchid Festival 2007 held at Perak, Malaysia. Presently she is working as Professor, Department of Pomology and Floriculture, College of Horticulture, Kerala Agricultural University.

FSC
www.fsc.org
MIX
Papier | Fördert gute Waldnutzung
FSC® C083411

Zeitfracht Medien GmbH
Ferdinand-Jühlke-Straße 7
99095 Erfurt, Deutschland
produktsicherheit@kolibri360.de